LECTURE NOTES

IN PURE AND APPLIED MATHEMATICS

1. *N. Jacobson,* Exceptional Lie Algebras
2. *L. -Å. Lindahl and F. Poulsen,* Thin Sets in Harmonic Analysis
3. *I. Satake,* Classification Theory of Semi-Simple Algebraic Groups
4. *F. Hirzebruch, W. D. Newmann, and S. S. Koh,* Differentiable Manifolds and Quadratic Forms (out of print)
5. *I. Chavel,* Riemannian Symmetric Spaces of Rank One (out of print)
6. *R. B. Burckel,* Characterization of C(X) Among Its Subalgebras
7. *B. R. McDonald, A. R. Magid, and K. C. Smith,* Ring Theory: Proceedings of the Oklahoma Conference
8. *Y.-T. Siu,* Techniques of Extension on Analytic Objects
9. *S. R. Caradus, W. E. Pfaffenberger, and B. Yood,* Calkin Algebras and Algebras of Operators on Banach Spaces
10. *E. O. Roxin, P.-T. Liu, and R. L. Sternberg,* Differential Games and Control Theory
11. *M. Orzech and C. Small,* The Brauer Group of Commutative Rings
12. *S. Thomeier,* Topology and Its Applications
13. *J. M. Lopez and K. A. Ross,* Sidon Sets
14. *W. W. Comfort and S. Negrepontis,* Continuous Pseudometrics
15. *K. McKennon and J. M. Robertson,* Locally Convex Spaces
16. *M. Carmeli and S. Malin,* Representations of the Rotation and Lorentz Groups: An Introduction
17. *G. B. Seligman,* Rational Methods in Lie Algebras
18. *D. G. de Figueiredo,* Functional Analysis: Proceedings of the Brazilian Mathematical Society Symposium
19. *L. Cesari, R. Kannan, and J. D. Schuur,* Nonlinear Functional Analysis and Differential Equations: Proceedings of the Michigan State University Conference
20. *J. J. Schäffer,* Geometry of Spheres in Normed Spaces
21. *K. Yano and M. Kon,* Anti-Invariant Submanifolds
22. *W. V. Vasconcelos,* The Rings of Dimension Two
23. *R. E. Chandler,* Hausdorff Compactifications
24. *S. P. Franklin and B. V. S. Thomas,* Topology: Proceedings of the Memphis State University Conference
25. *S. K. Jain,* Ring Theory: Proceedings of the Ohio University Conference
26. *B. R. McDonald and R. A. Morris,* Ring Theory II: Proceedings of the Second Oklahoma Conference
27. *R. B. Mura and A. Rhemtulla,* Orderable Groups
28. *J. R. Graef,* Stability of Dynamical Systems: Theory and Applications
29. *H.-C. Wang,* Homogeneous Branch Algebras
30. *E. O. Roxin, P.-T. Liu, and R. L. Sternberg,* Differential Games and Control Theory II
31. *R. D. Porter,* Introduction to Fibre Bundles
32. *M. Altman,* Contractors and Contractor Directions Theory and Applications
33. *J. S. Golan,* Decomposition and Dimension in Module Categories
34. *G. Fairweather,* Finite Element Galerkin Methods for Differential Equations
35. *J. D. Sally,* Numbers of Generators of Ideals in Local Rings
36. *S. S. Miller,* Complex Analysis: Proceedings of the S.U.N.Y. Brockport Conference
37. *R. Gordon,* Representation Theory of Algebras: Proceedings of the Philadelphia Conference
38. *M. Goto and F. D. Grosshans,* Semisimple Lie Algebras
39. *A. I. Arruda, N. C. A. da Costa, and R. Chuaqui,* Mathematical Logic: Proceedings of the First Brazilian Conference

40. *F. Van Oystaeyen,* Ring Theory: Proceedings of the 1977 Antwerp Conference
41. *F. Van Oystaeyen and A. Verschoren,* Reflectors and Localization: Application to Sheaf Theory
42. *M. Satyanarayana,* Positively Ordered Semigroups
43. *D. L. Russell,* Mathematics of Finite-Dimensional Control Systems
44. *P.-T. Liu and E. Roxin,* Differential Games and Control Theory III: Proceedings of the Third Kingston Conference, Part A
45. *A. Geramita and J. Seberry,* Orthogonal Designs: Quadratic Forms and Hadamard Matrices
46. *J. Cigler, V. Losert, and P. Michor,* Banach Modules and Functors on Categories of Banach Spaces
47. *P.-T. Liu and J. G. Sutinen,* Control Theory in Mathematical Economics: Proceedings of the Third Kingston Conference, Part B
48. *C. Byrnes,* Partial Differential Equations and Geometry
49. *G. Klambauer,* Problems and Propositions in Analysis
50. *J. Knopfmacher,* Analytic Arithmetic of Algebraic Function Fields
51. *F. Van Oystaeyen,* Ring Theory: Proceedings of the 1978 Antwerp Conference
52. *B. Kedem,* Binary Time Series
53. *J. Barros-Neto and R. A. Artino,* Hypoelliptic Boundary-Value Problems
54. *R. L. Sternberg, A. J. Kalinowski, and J. S. Papadakis,* Nonlinear Partial Differential Equations in Engineering and Applied Science
55. *B. R. McDonald,* Ring Theory and Algebra III: Proceedings of the Third Oklahoma Conference
56. *J. S. Golan,* Structure Sheaves over a Noncommutative Ring
57. *T. V. Narayana, J. G. Williams, and R. M. Mathsen,* Combinatorics, Representation Theory and Statistical Methods in Groups: YOUNG DAY Proceedings
58. *T. A. Burton,* Modeling and Differential Equations in Biology
59. *K. H. Kim and F. W. Roush,* Introduction to Mathematical Consensus Theory
60. *J. Banas and K. Goebel,* Measures of Noncompactness in Banach Spaces
61. *O. A. Nielson,* Direct Integral Theory
62. *J. E. Smith, G. O. Kenny, and R. N. Ball,* Ordered Groups: Proceedings of the Boise State Conference
63. *J. Cronin,* Mathematics of Cell Electrophysiology
64. *J. W. Brewer,* Power Series Over Commutative Rings
65. *P. K. Kamthan and M. Gupta,* Sequence Spaces and Series
66. *T. G. McLaughlin,* Regressive Sets and the Theory of Isols
67. *T. L. Herdman, S. M. Rankin, III, and H. W. Stech,* Integral and Functional Differential Equations
68. *R. Draper,* Commutative Algebra: Analytic Methods
69. *W. G. McKay and J. Patera,* Tables of Dimensions, Indices, and Branching Rules for Representations of Simple Lie Algebras
70. *R. L. Devaney and Z. H. Nitecki,* Classical Mechanics and Dynamical Systems
71. *J. Van Geel,* Places and Valuations in Noncommutative Ring Theory
72. *C. Faith,* Injective Modules and Injective Quotient Rings
73. *A. Fiacco,* Mathematical Programming with Data Perturbations I
74. *P. Schultz, C. Praeger, and R. Sullivan,* Algebraic Structures and Applications Proceedings of the First Western Australian Conference on Algebra
75. *L. Bican, T. Kepka, and P. Nemec,* Rings, Modules, and Preradicals
76. *D. C. Kay and M. Breen,* Convexity and Related Combinatorial Geometry: Proceedings of the Second University of Oklahoma Conference
77. *P. Fletcher and W. F. Lindgren,* Quasi-Uniform Spaces
78. *C.-C. Yang,* Factorization Theory of Meromorphic Functions
79. *O. Taussky,* Ternary Quadratic Forms and Norms
80. *S. P. Singh and J. H. Burry,* Nonlinear Analysis and Applications
81. *K. B. Hannsgen, T. L. Herdman, H. W. Stech, and R. L. Wheeler,* Volterra and Functional Differential Equations

NONLINEAR FUNCTIONAL ANALYSIS

82. *N. L. Johnson, M. J. Kallaher, and C. T. Long,* Finite Geometries: Proceedings of a Conference in Honor of T. G. Ostrom
83. *G. I. Zapata,* Functional Analysis, Holomorphy, and Approximation Theory
84. *S. Greco and G. Valla,* Commutative Algebra: Proceedings of the Trento Conference
85. *A. V. Fiacco,* Mathematical Programming with Data Perturbations II
86. *J.-B. Hiriart-Urruty, W. Oettli, and J. Stoer,* Optimization: Theory and Algorithms
87. *A. Figa Talamanca and M. A. Picardello,* Harmonic Analysis on Free Groups
88. *M. Harada,* Factor Categories with Applications to Direct Decomposition of Modules
89. *V. I. Istrăţescu,* Strict Convexity and Complex Strict Convexity: Theory and Applications
90. *V. Lakshmikantham,* Trends in Theory and Practice of Nonlinear Differential Equations
91. *H. L. Manocha and J. B. Srivastava,* Algebra and Its Applications
92. *D. V. Chudnovsky and G. V. Chudnovsky,* Classical and Quantum Models and Arithmetic Problems
93. *J. W. Longley,* Least Squares Computations Using Orthogonalization Methods
94. *L. P. de Alcantara,* Mathematical Logic and Formal Systems
95. *C. E. Aull,* Rings of Continuous Functions
96. *R. Chuaqui,* Analysis, Geometry, and Probability
97. *L. Fuchs and L. Salce,* Modules Over Valuation Domains
98. *P. Fischer and W. R. Smith,* Chaos, Fractals, and Dynamics
99. *W. B. Powell and C. Tsinakis,* Ordered Algebraic Structures
100. *G. M. Rassias and T. M. Rassias,* Differential Geometry, Calculus of Variations, and Their Applications
101. *R.-E. Hoffmann and K. H. Hofmann,* Continuous Lattices and Their Applications
102. *J. H. Lightbourne, III, and S. M. Rankin, III,* Physical Mathematics and Nonlinear Partial Differential Equations
103. *C. A. Baker and L. M. Batten,* Finite Geometries
104. *J. W. Brewer, J. W. Bunce, and F. S. Van Vleck,* Linear Systems Over Commutative Rings
105. *C. McCrory and T. Shifrin,* Geometry and Topology: Manifolds, Varieties, and Knots
106. *D. W. Kueker, E. G. K. Lopez-Escobar, and C. H. Smith,* Mathematical Logic and Theoretical Computer Science
107. *B.-L. Lin and S. Simons,* Nonlinear and Convex Analysis: Proceedings in Honor of Ky Fan
108. *S. J. Lee,* Operator Methods for Optimal Control Problems
109. *V. Lakshmikantham,* Nonlinear Analysis and Applications
110. *S. F. McCormick,* Multigrid Methods: Theory, Applications, and Supercomputing
111. *M. C. Tangora,* Computers in Algebra
112. *D. V. Chudnovsky and G. V. Chudnovsky,* Search Theory: Some Recent Developments
113. *D. V. Chudnovsky and R. D. Jenks,* Computer Algebra
114. *M. C. Tangora,* Computers in Geometry and Topology
115. *P. Nelson, V. Faber, T. A. Manteuffel, D. L. Seth, and A. B. White, Jr.* Transport Theory, Invariant Imbedding, and Integral Equations: Proceedings in Honor of G. M. Wing's 65th Birthday
116. *P. Clément, S. Invernizzi, E. Mitidieri, and I. I. Vrabie,* Semigroup Theory and Applications
117. *J. Vinuesa,* Orthogonal Polynomials and Their Applications: Proceedings of the International Congress
118. *C. M. Dafermos, G. Ladas, and G. Papanicolaou,* Differential Equations: Proceedings of the EQUADIFF Conference
119. *E. O. Roxin,* Modern Optimal Control: A Conference in Honor of Solomon Lefschetz and Joseph P. Lasalle
120. *J. C. Díaz,* Mathematics for Large Scale Computing
121. *P. S. Milojević,* Nonlinear Functional Analysis

Other Volumes in Preparation

NONLINEAR FUNCTIONAL ANALYSIS

Edited by

P. S. Milojević
New Jersey Institute of Technology
Newark, New Jersey

MARCEL DEKKER, INC. New York and Basel

Library of Congress Cataloging-in-Publication Data

Nonlinear functional analysis / edited by P. S. Milojević.
p. cm. -- (Pure and applied mathematics ; 121)
'Papers . . . based on the lectures presented at the Special Session on Nonlinear Functional Analysis of the American Mathematical Society Regional Meeting held at New Jersey Institute of Technology in Newark, New Jersey, on April 25-26, 1987'--Pref.
ISBN 0-8247-8255-0 (alk. paper)
1. Nonlinear functional analysis--Congresses. I. Milojević, P. S., II. Special Session on Nonlinear Functional Analysis (1987 : Newark, N.J.) III. Series.
QA321.5.N662 1989
515'.7--dc20 89-17173
CIP

This book is printed on acid-free paper

MARCEL DEKKER, INC.
270 Madison Avenue, New York, New York 10016

Current printing (last digit):
10 9 8 7 6 5 4 3 2 1

PRINTED IN THE UNITED STATES OF AMERICA

Preface

These proceedings consist of a collection of papers, most of which are based on the lectures presented at the Special Session on Nonlinear Functional Analysis of the American Mathematical Society Regional Meeting, held at New Jersey Institute of Technology in Newark, New Jersey, on April 25–26, 1987. The papers of C. Viterbo and K. Wysocki were contributed by invitation, while the paper by G. Fournier and M. Martelli covers only one of the two topics presented by the second author.

The topics covered in the proceedings are global invertibility and finite solvability of nonlinear differential equations, existence of positive eigenvalues and eigenvalues for compact and condensing-like maps, recent progress in Hamiltonian dynamics and symplectic geometry, solvability of semilinear operator and hyperbolic differential equations, operator-valued means and their iterates, and existence of multiple critical points of functionals preserving an order structure.

I thank all the contributors for making these proceedings possible.

P. S. Milojević

Contents

Contributors

V. CAFAGNA* Université Paris Nord, C. S. P., Département de Mathématiques et d'Informatique, Villetaneuse, France

G. FOURNIER Département de Mathématique, Université de Sherbrooke, Sherbrooke, Quebec, Canada

H. HOFER Department of Mathematics, Rutgers University, New Brunswick, New Jersey

M. MARTELLI Department of Mathematics, California State University, Fullerton, California

P. S. MILOJEVIĆ Department of Mathematics, New Jersey Institute of Technology, Newark, New Jersey

ROGER D. NUSSBAUM Department of Mathematics, Rutgers University, New Brunswick, New Jersey

CLAUDE VITERBO Ceremade, Université de Paris, Paris, France, and Mathematical Sciences Research Institute, Berkeley, California

KRZYSZTOF WYSOCKI† Department of Mathematics, Rutgers University, New Brunswick, New Jersey

---

*Permanent address: Istituto di Matematica, Facoltà di Scienze, Università di Salerno, Italy.

†Current affiliation: Center for Dynamical Systems and Nonlinear Studies, School of Mathematics, Georgia Institute of Technology, Atlanta, Georgia

NONLINEAR FUNCTIONAL ANALYSIS

Global Invertibility and Finite Solvability

V. CAFAGNA* Université Paris Nord, C.S.P., Département de Mathématiques et d'Informatique, 93430 Villetaneuse, FRANCE.

INTRODUCTION

Aim of this paper is to describe some aspects of the map approach to nonlinear differential equations. These will be illustrated by means of a single class of examples, namely semilinear Dirichlet problems :

$$\begin{cases} \Delta u + f(u) = h & \text{in} \quad \Omega \\ u = 0 & \text{on} \quad \partial\Omega \ . \end{cases} \tag{1}$$

The adopted point of view consists in studying the global features of the map

$$F : C_o^{2,\alpha}(\overline{\Omega}) \to C^{o,\alpha}(\overline{\Omega}) \tag{2}$$

defined by $F(u) = \Delta u + f(u)$. Then the solutions of problem (1) are described as points of the preimage set $F^{-1}(h)$. This is, so to speak, in contrast with the variational method, which describes the solutions of problem (1) as critical points of the functional

$$\Gamma(u) = \int_\Omega \left(\frac{1}{2} |\nabla u|^2 - F(u) - hu\right) dx \tag{3}$$

See Vainberg (1968) for a comparative illustration of both methods.

Obviously the Leray-Schauder degree theory is the best known

* Permanent address : Istituto di Matematica, Facoltà di Scienze, Università di Salerno, 84100 Salerno, Italy.

example of the map point of view, while here we stress a slightly different appraoch which is essentially due to Caccioppoli (1932). The main point in the Caccioppoli method is that global features of the map F are exploited in order to try to "invert" it in some suitable sense. This permits, in some situations, to completely determine the number of solutions to problem (1).

The nicest behaviour one can expect from a map is the absence of singularities, so that it behaves locally as a diffeomorphism. If this nice local behaviour is combined with a nice behaviour at infinity (i.e. properness), then the map is a global diffeomorphism and problem (1), which the map represents, has one and only one solution for each right-hand side. This is *grosso modo* the content of the Global Invertibility Theorem.

But Caccioppoli's analysis was not limited to non-singular maps. In a paper of 1936 he provided local analyses of singularities, proving a version of what is nowadays generally known as the Local Representative Theorem of Fredholm maps (see Abraham and Robbin (1972)). So, in a sense, Caccioppoli can be considered as the father (or maybe the grandfather) of the modern singularity theory of Fredholm maps.

In the seventies Ambrosetti and Prodi (1972) returned to the ideas of Caccioppoli on global invertibility and were able to give a full description of the behaviour of a map with singularities, defined by a semilinear Dirichlet problem like (1), with suitable hypotheses on f. Nowadays their result is understood as being an example of an infinite-dimensional analog of the Whitney Fold (Berger and Church (1979), Cafagna (1984)).

Use of differential-topological methods coming from the singularity theory of smooth maps revitalized the subject in recent years. Higher order singularities have been investigated and many new applications have been given, beside elliptic equations (Berger, Church and Timourian (1987), Lazzeri and Micheletti (1988), Cafagna and Tarantello (1987)), to different problems as, for example, periodic solutions of first order differential equations (McKean and Scovel (1986), Cafagna and Donati (1985)), and forced oscillations of conservative systems (Damon (1987), Cafagna (1986)). In particular cusp singularities have been object of intensive studies and some examples of infinite-dimensional analogs of the Whitney Cusp have been discovered (Cafagna and Donati (1985), Ruf (1988)), thus extending the list of globally invertible maps (with singularities), defined by differential equations.

In all the papers quoted above the accent is mainly on the full description of the behaviour of the map in the neighborhood of a singularity. So the examples treated with these methods are necessarily very simple in nature. In fact things get immediately more and more complicated as soon as the hypotheses on the nonlinearity are a little bit relaxed. Even a rather innocent-looking problem like

$$\begin{cases} u'' + u^2 = h & \text{in } (0,\pi) \\ u(0) = u(\pi) = 0 \end{cases} \tag{4}$$

gives rise to a map $F : C_o^2([0,\pi]) \to C^o([0,\pi])$ for which one can show (McKean (1987)) that every kind of singularity of the Morin list can occur (see Morin (1965) and also Golubitsky and Guillemin (1973) for the full list of the Morin singularities). So there exist right-hand sides with an arbitrary large set of solutions. Nontheless one can prove (McKean and Scovel (1986)) that the solution set is always finite. In some sense this finite solvability property is the most general phenomenon of global invertibility. The "inverse" of a finitely solvable map being a collection of multi-valued maps.

After having discussed invertible maps without singularities and "invertible" maps with simple singularities (folds and cusps) we sketch in the last section of the paper a theory of finite solvability for a class of real-analytic Fredholm maps. Some examples (taken from Cafagna and Donati (1986 and 1988a)) are discussed.

The style of this exposition is informal, proofs and theorems are usually only sketched and in general the accent is more on the geometrical than the analytical side of the subject , more on singularity theory than on *a priori* estimates.

1 GLOBAL INVERTIBILITY

Let us begin by recalling some simple facts of functional analysis. Let X and Y be Banach spaces and $F : X \to Y$ be a smooth map. Needless to say some finite order of differentiability would be enough for almost all the considerations in this paper, but we prefer to be imprecise rather than pedantic. By $F' : X \to L(X,Y)$ we intend the Fréchet derivative of the map F. Higher order derivatives will be indicated by the symbol $F^{(k)}$. Recall that the k-th derivative of F is a map $F^{(k)} : X \to L^k(X,Y)$ from X to the space of k-linear maps of X into Y.

We say that a linear operator $A \in L(X,Y)$ is *Fredholm* if a) Ker A is finite-dimensional ; b) Im A is closed in Y and finite-codimensional. The *index* of A is defined by $i_A = \dim \operatorname{Ker} A - \operatorname{codim} \operatorname{Im} A$. We say that a smooth map $F : X \to Y$ is a *Fredholm map* if $F'(u) \in L(X,Y)$ is a Fredholm operator for any $u \in X$. It is easy to show that the index $i_{F'(u)}$ does not depend on u, so that we may speak of the *index* of F.

In this paper we will only be concerned with Fredholm maps of index 0. Let $F : X \to Y$ be such a map. We say that $u \in X$ is a *singular point* of F (or a *singularity*) if $\operatorname{Ker} F'(u)$ contains a nontrivial vector (equivalently, if $\dim \operatorname{Ker} F'(u) \geq 1$). Otherwise u is said to be a *regular point*. We will usually indicate by S_F (or simply S when no confusion is possible) the *singular set* of F :

$$S_F = \{u \in X ; \dim \operatorname{Ker} F'(u) \geq 1\} \tag{5}$$

A point $h \in Y$ is called a *regular value* of F, if it is not contained in the image $F(S)$ of the singular set. Otherwise it is called a *critical value*. Let $u \in X$ be a regular point of F ; then the Inverse Function Theorem says that the map F behaves locally like the identity. To be more precise one can restate the theorem in the following rather involved, but pedagogically useful, way :

INVERSE FUNCTION THEOREM. Let X and Y be Banach spaces and $F : X \to Y$ be a smooth Fredholm map of index 0. Let u be a regular point of F; then there exist a neighborhood U of u in X, a neighborhood V of $F(u)$ in Y, a Banach space Z and diffeomorphisms $\phi : U \to Z$, $\psi : V \to Z$ such that $(\psi \circ F \circ \phi^{-1})(z) = I(z) = z$. That is to say the following diagram commutes

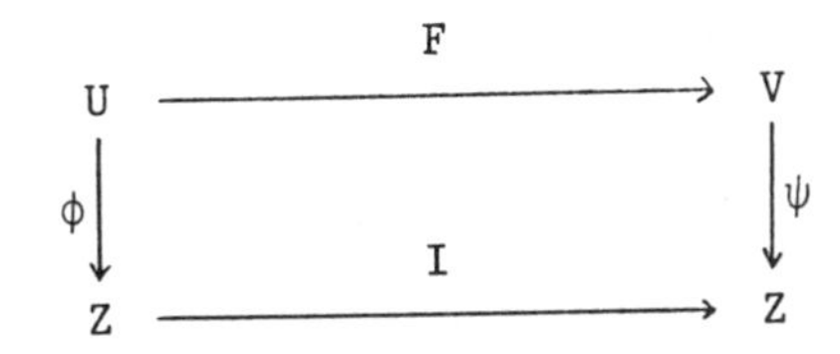

I is called the *local representative* of F.

In the next section we will analyze the behaviour of maps near some simple singularities and state some normal form theorems which might be regarded as generalizations of the Inverse Function Theorem.

We are now in a position to state the Global Invertibility Theorem. Just recall first that a map $F : X \to Y$ is said to be *proper* if the inverse image of every compact subset of Y is compact in X.

GLOBAL INVERSION THEOREM. Let $F : X \to Y$ be a proper smooth Fredholm map of index 0. If the singular set of F is empty, then F is a diffeomorphism of X onto Y.

Remark. The theorem is apparently due independently to Caccioppoli (1932) and Banach and Mazur (1934).

Sketch of the proof. (For the missing details we refer to Plastock (1974), Berger (1977) and Prodi and Ambrosetti (1973)). Let $h \in Y$. Then $F^{-1}(h)$ is a finite set, being compact, by the properness, and discrete, by the Inverse Function Theorem. It is also easy to see that $F^{-1}(h)$ contains the same number of points for every $h \in Y$: just define a smooth arc γ between any couple of values h_1 and h_2 in Y ; then properness and regularity imply that $F^{-1}(\gamma)$ is a finite set of arcs connecting points in $F^{-1}(h_1)$ to points in $F^{-1}(h_2)$. So F is a finitely-sheeted covering map of X onto Y . It is then an easy corollary to the Covering Homotopy Theorem (see Greenberg (1966)) that a finite covering of a simply connected space (what is obviously the case of the vector space Y) must have only one sheet. This amounts to say that F has to be 1-1 .

Now for completeness we are going to give a sketchy treatment of a classical application of this theorem to an existence and uniqueness result for a semilinear Dirichlet problem.

Let $\Omega \subset \mathbb{R}^n$ be a bounded domain with smooth boundary $\partial\Omega$. Consider the eigenvalue problem

$$\begin{cases} \Delta v + \lambda v = 0 & \text{in} \quad \Omega \\ v = 0 & \text{on} \quad \partial\Omega . \end{cases} \tag{6}$$

Denote by $0 < \lambda_1 < \lambda_2 \leqq \dots \leqq \lambda_k \leqq \dots$ the sequence of eigenvalues and by ϕ_k the corresponding eigenfunctions (normalized so that $\int_\Omega |\phi_k(x)|^2 dx = 1$). We also need to consider the following slightly more general kind of eigenvalue problem with a weight

$$\begin{cases} \Delta v + \mu P V = 0 & \text{in} \quad \Omega \\ v = 0 & \text{on} \quad \partial\Omega \end{cases} \tag{7}$$

Here $P : \Omega \to \mathbb{R}$ is a smooth positive function. One has the following

COMPARISON CRITERIA. The eigenvalues $\mu_j = \mu_j(P)$ of problem (7) form a positive increasing sequence. μ_1 is simple and the corresponding eigenfunctions do not change sign. If $P < Q$ are two different weights, then $\mu_j(P) > \mu_j(Q)$ for every $j \in \mathbb{N}$.

For a proof see Courant and Hilbert (1953).

Let now $C^{k,\alpha}(\overline{\Omega})$ be the space of k times differentiable functions on $\overline{\Omega}$, whose k -th derivatives are Hölder continuous with exponent α and $C_o^{2,\alpha}(\overline{\Omega}) = \{u \in C^{2,\alpha}(\overline{\Omega}) : u = 0 \text{ on } \partial\Omega\}$. Given $h \in C^{o,\alpha}(\overline{\Omega})$ we will look for functions $u \in C_o^{2,\alpha}(\overline{\Omega})$ which solve the Dirichlet problem

$$\begin{cases} \Delta u + f(u) = h & \text{in } \Omega \\ u = 0 & \text{on } \partial\Omega \end{cases} \tag{8}$$

As an application of the global invertibility machinery we will give a proof of the following classical result (Caccioppoli (1932). See also Hammerstein (1929) and Dolph (1949)).

THEOREM 1. Let $\ell, m \in \mathbb{R}$ be such that

$$\lambda_k < \ell \leqq f'(x) \leqq m \leqq \lambda_{k+1} \quad , \quad \forall\, x \in \overline{\Omega} \tag{9}$$

Then for every $h \in C^{o,\alpha}(\overline{\Omega})$ there exists a unique $u \in C_o^{2,\alpha}(\overline{\Omega})$ solution of (8).

Proof. Consider the map $F : C_o^{2,\alpha}(\overline{\Omega}) \to C^{o,\alpha}(\overline{\Omega})$ defined by $F(u) = \Delta u + f(u)$. We will prove that for every u the Fréchet derivative F' is a linear isomorphism of $C_o^{2,\alpha}(\overline{\Omega})$ onto $C^{o,\alpha}(\overline{\Omega})$. This will show that F is a Fredholm map of index 0 without singularities. In fact u is a singularity of F if and only if there are nontrivial solutions to the linearized problem

$$\begin{cases} \Delta v + f'(u)v = 0 & \text{in } \Omega \\ v = 0 & \text{on } \partial\Omega \end{cases} \quad , \tag{10}$$

that is to say if and only if some eigenvalue $\mu_j(f'(u))$ of the eigenvalue problem (7) with weight $P = f'(u)$ is equal to 1. But this is absurd, because from the Comparison Criteria one obtains

$$\frac{\mu_j}{\mu_{k+1}} = \mu_j(\lambda_{k+1}) < \mu_j(f'(u)) < \mu_j(\lambda_k) = \frac{\lambda_j}{\lambda_k} \tag{11}$$

and this readily denies any possibility for $\mu_j(f'(u))$ of being equal to 1 .

In order to prove the properness of the map F one has to show that any sequence $\{u_n\} \subset C^{2,\alpha}(\overline{\Omega})$ such that $F(u_n)$ is convergent, has a convergent subsequence. It is easy to see that it is enough to show that u_n is actually bounded in $C^{o,\alpha}(\overline{\Omega})$. In fact in this case $\Delta u_n = h_n - f(u_n)$ is also bounded in $C^{o,\alpha}(\overline{\Omega})$ and then u_n is bounded in $C^{2,\alpha}(\overline{\Omega})$ (being $\Delta : C_o^{2,\alpha}(\overline{\Omega}) \to C^{o,\alpha}(\overline{\Omega})$ an isomorphism). Then (a subsequence of) u_n converges to u in $C^{1,\alpha}(\overline{\Omega})$, and so Δu_n converges in $C^{o,\alpha}(\overline{\Omega})$ and u_n converges in $C^{2,\alpha}(\overline{\Omega})$.

We will now prove $C^{o,\alpha}$ -boundedness of u_n . Suppose that $\|u_n\|_{o,\alpha} \to +\infty$. Define a sequence z_n on the unit sphere of $C^{2,\alpha}(\overline{\Omega})$ by $z_n = u_n(\|u_n\|_{o,\alpha})^{-1}$. The functions z_n satisfy

$$\Delta z_n + \omega(u_n)z_n = \frac{h_n}{\|u_n\|_{o,\alpha}} , \tag{12}$$

where $\omega : \mathbb{R} \to \mathbb{R}$ is defined by

$$\omega(t) = \begin{cases} \dfrac{f(t)}{t} & t \neq 0 \\ f'(0) & t = 0 \end{cases} . \tag{13}$$

From linear elliptic theory one deduces that a subsequence of z_n converges to a function z^* living on the unit sphere of $C_o^1(\overline{\Omega})$. Taking the limit as $n \to +\infty$ in equation (10) one sees that z^* is a solution of

$$\begin{cases} \Delta z^* + \mu\omega^* z^* = 0 & \text{in} \quad \Omega \\ z^* = 0 & \text{on} \quad \partial\Omega , \end{cases} \tag{14}$$

where ω^* is the weak limit (in $L^1(\Omega)^*$) of the $\omega(u_n)$. But this contradicts the Comparison Criteria, because ω^* satisfies

$$\lambda_k < \ell \leq \omega^* \leq m < \ell_{k+1} . \tag{15}$$

2 MAPS WITH SIMPLE SINGULARITIES.

In the last paragraph we considered the case of a Dirichlet problem whose nonlinearity has no interaction with the spectrum of the "linear part" ($-\Delta$, to be precise). This situation gives rise to a map without singularities. Now we want to investigate what happens when singularities come into play. We will show that in some simple cases one is still able to give a complete description of the behaviour of the map (at least as far as the number of solutions is concerned). In some sense one will be still able to define an "inverse" of the map, this time not a true function, but rather a collection of multi-valued maps.

There is a natural hierarchy of the singularities of a map with respect to the simplicity of the behaviour of the map nearby, the simplest behaviour being obviously no singularity at all, i.e. a local diffeomorphism. We will only discuss the first two steps of this hierarchy, namely fold and cusp singularities. Moreover we will take into consideration only maps arising from nonlinear Dirichlet problems (and needless to say we give only a sketchy description of the theory). For the general case, as well as for complete proofs we refer to Cafagna and Donati (1988b) (see also Berger, Church and Timourian (1985), Damon (1986) and Lazzeri and Micheletti (1988)).

We will need in the sequel some more nonlinear functional analysis. Let M and N be a Banach manifolds and $Z \subset N$ a smooth finite-codimensional submanifold of N . Let $y \in Z$, by $T_y Z$ we mean the tangent space to Z at y . We say that a smooth map $G : M \to N$ is *transverse to* Z *at* $u \in M$, and we write $G \pitchfork_u Z$, if either $G(u) \notin Z$ or $G(u) \in Z$ and

$$G'(u)T_u M + T_{G(u)} Z = T_{G(u)} N \quad . \tag{16}$$

If $G \pitchfork_u Z$ for every $u \in M$, we say that G is *transverse to* Z and write $G \pitchfork Z$. One of the essential features of transverse maps is the nice behaviour with respect to preimages captured in the following :

INVERSE IMAGE THEOREM. Let M and N be Banach manifolds, $Z \subset N$ a submanifold of N of codimension k and $G : M \to N$ a smooth map transverse to Z. Then $G^{-1}(Z)$, if nonempty, is a smooth submanifold of M of codimension k. Moreover one has the following formula for the tangent space to $G^{-1}(z)$ at u:

$$T_u G^{-1}(z) = [G'(u)]^{-1} T_{G(u)} Z \tag{17}$$

For a proof we refer to Borisovich, Zvyagin and Sapronov (1977). Let us just remark that in the case $N = \mathbb{R}$ and $Z = \{y\}$, the condition $G \pitchfork \{y\}$ means nothing else but that y is a regular value of G. In this case the above theorem says that inverse images of regular values of real-valued functions are smooth hypersurfaces.

As an almost immediate application of this theorem we give a description of the structure of the singular set of a map defined by a Dirichlet problem with a nonlinearity interacting in a nice way only with a simple eigenvalue.

THEOREM 2. Let $\Omega \subset \mathbb{R}^n$ be a bounded domain with smooth boundary, and λ_k be a simple eigenvalue of the eigenvalue problem (6). Let $f \in C^\infty(\mathbb{R})$ be such that

$$\begin{aligned} &\text{i)} \quad \lambda_{k-1} < f'(x) < \lambda_{k+1} \qquad \forall\, x \in \overline{\Omega} \\ &\text{ii)} \quad \exists\, \underline{x},\ \bar{x} \in \Omega : \quad f'(\underline{x}) < \lambda_k < f'(\bar{x}) \\ &\text{iii)} \quad \exists\, \varepsilon > 0 : \quad f''(x) \not\equiv 0 \quad \text{in } (-\varepsilon, \varepsilon) \end{aligned} \tag{18}$$

Then the singular set of the map $F : C_o^{2,\alpha}(\overline{\Omega}) \to C^{o,\alpha}(\overline{\Omega})$ defined by $F(u) = \Delta u + f(u)$ is a smooth hypersurface.

Proof. Being λ_k simple, the Comparison Criteria imply that

$$\dim \operatorname{Ker} F'(u) = 1 \qquad \forall\, u \in S_F \tag{19}$$

To simplify the notations, from now on, we will write X for $C_o^{2,\alpha}(\overline{\Omega})$ and Y for $C^{o,\alpha}(\overline{\Omega})$. Consider then in the space $L(X,Y)$ the subset Σ of Fredholm operators of index 0 with one-dimensional kernel. One can show (see for example Cafagna and Donati (1988b)) that Σ is a smooth hypersurface of $L(X,Y)$ and that the tangent hyperplane to Σ at a point A is given by

$$T_A\Sigma = \{B \in L(X,Y) \; ; \; B \text{ Ker } A \subset \text{Im } A\} \tag{20}$$

Now by (19) it is immediate that $u \in S_F$ if and only if $F'(u) \in \Sigma$, i.e.

$$S_F = (F')^{-1} \Sigma \quad . \tag{21}$$

We note that hypothesis (18) ii) implies that S is actually non empty. So in order to conclude, by the Inverse Image Theorem, it is sufficient to show that the Fréchet derivative $F' : X \to L(X,Y)$ is transverse to Σ . The transversality condition (16) reads in this case

$$F''(u)X + T_{F'(u)}\Sigma = T_{F'(u)}L(X,Y) \equiv L(X,Y) \qquad \forall\, u \in S \tag{22}$$

Being $T_{F'(u)}\Sigma$ a hyperplane (22) is verified once one shows the existence of a $v \in X$ such that

$$F''(u)v \notin T_{F'(u)}\Sigma \tag{23}$$

Recalling the formula (20) for $T_{F'(u)}\Sigma$, this is seen to be equivalent to the existence of a $v \in X$ such that

$$F'(u)(v,\phi) \notin \text{Im } F'(u) \tag{24}$$

where ϕ is any nontrivial function in Ker $F'(u)$. Remark that, by the autoadjoint character of the Dirichlet problem, one has

$$\text{Im } F'(u) = \left\{h \in Y : \int_\Omega \phi h \, dx = 0\right\} \quad , \quad (\phi \in \text{Ker } F'(u)) \; . \tag{25}$$

So the transversality condition finally becomes

$$\exists\, v \in X : \quad \int_\Omega f''(u) v\, \phi^2\, dx \neq 0 \tag{26}$$

which is trivially verified, by hypothesis (18) iii).

From now on we will call 1 *-transverse at* $u \in S_F$ a Fredholm map F of index 0 such that $\dim \operatorname{Ker} F'(u) = 1$ and such that the transversality hypothesis (22) is verified at u. Sometimes we will also say by *abus de langage* that u is a 1 *-transverse singularity*. A map 1 -transverse at every $u \in S$ will be shortly called 1 *-transverse*. What we have really proved in Theorem 2 is : 1. The singular set of a 1 -transverse map is a smooth hypersurface. 2. The map defined by the conditions (18) is 1 -transverse.

We say that a 1 -transverse singularity u of a map F is a *fold* if $\operatorname{Ker} F'(u)$ is not tangent to S. Recalling the formula (17) one has that

$$\begin{aligned} T_u S &= [F''(u)]^{-1}\, T_{F(u)}\Sigma \\ &= \{v \in X : F''(u)(v,\phi) \notin \operatorname{Im} F'(u)\} \end{aligned} \tag{27}$$

where, as usual, ϕ is a nontrivial function in $\operatorname{Ker} F'(u)$. Then the condition for $u \in S$ to be a fold becomes

$$F''(u)(u,\phi) \notin \operatorname{Im} F'(u) \tag{28}$$

When $F = \Delta + f$, this condition becomes

$$\int_\Omega f''(u)\phi^3\, dx \neq 0 \quad . \tag{29}$$

The behaviour of a map in a neighborhood of a fold is in some sense the simplest among all the singular behaviours, namely the map is like the identity on a hyperplane and behaves like the function $f(t) = t^2$ in the left-out direction, as one can see in the following :

NORMAL FORM THEOREM FOR FOLDS. Let u be a fold singularity of the map $F : X \to Y$. Then there exist a neighborhood U of u in X, a neighborhood V of $F(u)$ in Y, a Banach space Z and diffeomorphisms $\phi : U \to \mathbb{R} \times Z$, $\psi : V \to \mathbb{R} \times Z$ such that $(\psi \circ F \circ \phi^{-1})(t,z) = \tilde{F}(t,z) = (t^2,z)$. That is to say, the following diagram commutes

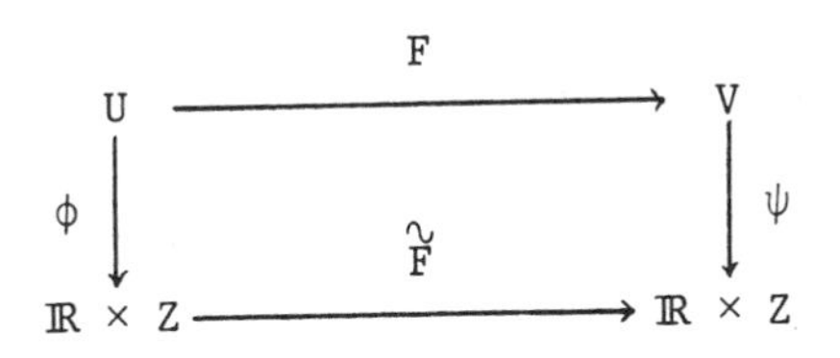

A proof of this theorem can be found in any book on singularity theory (for example Golubitsky and Guillemin (1973) and Martinet (1982)), where it is usually a particular case of the more general normal form theorem for the Morin singularities. A simple proof is reported in Cafagna (1984).

Before we continue our brief *excursus* of singularity theory, we would like to sketch a proof of an influential theorem by Ambrosetti and Prodi (1972).

THEOREM 3. Let $\Omega \subset \mathbb{R}^n$ be a bounded domain with smooth boundary. Let $f \in C^\infty(\mathbb{R})$ be such that

$$\begin{array}{ll} \text{i)} & \lim_{t \to -\infty} f'(t) < \lambda_1 < \lim_{t \to +\infty} f'(t) < \lambda_2 \\ \text{ii)} & f''(t) > 0 \qquad \forall\, t \in \mathbb{R} \end{array} \tag{30}$$

and let $F : C_0^{2,\alpha}(\overline{\Omega}) \to C^{0,\alpha}(\overline{\Omega})$ be defined by $F(u) = \Delta u + f(u)$. Then there is a smooth hypersurface Z which disconnects $C^{0,\alpha}(\overline{\Omega})$ in two components Y_2 and Y_0 such that

$$\begin{array}{lll} F^{-1}(h) = \emptyset & & \forall\, h \in Y_0 \\ \# F^{-1}(h) = 2 & & \forall\, h \in Y_2 \\ \# F^{-1}(h) = 1 & & \forall\, h \in Z \end{array} \tag{31}$$

(here "$\sharp$" denotes cardinality of a set).

Sketch of the proof. Hypotheses (30) imply that F is a 1-transverse map, so that S is a smooth hypersurface. We will prove that $Z = F(S)$ does the job.

First of all we remark that every singularity is a fold. In fact, by the Comparison Criteria, one has that any function $\phi \in \operatorname{Ker} F'(u)$ is of one sign. This and the strict convexity of f imply that the fold condition (29)

$$\int_{\Omega} f''(u)\phi^3 \, dx \neq 0$$

is verified $\forall\, u \in S$.

It is also easy to see (using the Comparison Criteria once more) that F is 1-1 on S. In fact suppose by absurd the existence of two different functions $u,v \in S$ with $F(u) = F(v)$. Then $(u-v)$ would be an eigenfunction relative to the eigenvalue $\mu_1 = 1$ for the weight ω defined by

$$\omega(x) = \begin{cases} \dfrac{f(u(x)) - f(v(x))}{u(x) - v(x)} & \forall\, x : u(x) \neq v(x) \\ f'(u(x)) & \forall\, x : u(x) = v(x) \end{cases} \tag{32}$$

Being u singular, 1 will also be the first eigenvalue for the weight function $f'(u)$. But convexity of f implies that $f'(u) < \omega$ and this is in contradiction with the Comparison Criteria.

After that one can show that the map F is proper by proving an *a priori* estimate similar to the one in Theorem 1.

Then one concludes that $F(S) = Z$ is a smooth hypersurface of $C^{0,\alpha}(\overline{\Omega})$ diffeomorphic to S, and the thesis follows by applying the Normal Form Theorem for Folds.

The geometrical meaning of all this is that the map defined by the above theorem behaves globally like the Whitney Fold map $\mathbb{R}^2 \to \mathbb{R}^2$ defined by $(t,s) \to (t^2,s)$ (Whitney (1955)). In fact Berger and Church (1979) proved that there exists actually a global coordinate change which send F into the above normal form.

Other applications of the study of fold singularities to different nonlinear problems are known. In particular the papers of McKean and Scovel (1986) and Cafagna and Donati (1988b) contain a study of periodic solutions of the Riccati equation *via* singularity theory.

The next step in singularity analysis is the consideration of cusps. In developing a theory of cusps we will be considerably less precise. Let $F : X \to Y$ be a Fredholm map of index 0 and let u be a 1-transverse singularity of F. Suppose that u is not a fold, i.e. $\text{Ker } F'(u) < T_u S$. Suppose also that F verifies some higher order transversality condition (which we won't specify further) called $(1,1)$-transversality implying that the set of 1-transverse singularities which are not folds is (in a neighborhood of u) a smooth hypersurface $S_{1,1}$ of the singular hypersurface S. If $\text{Ker } F'(u)$ is not tangent to $S_{1,1}$ we say that u is a *cusp*.

For maps arising from Dirichlet problems we have the following characterizations of $(1,1)$-transverse and cusp singularities :

PROPOSITION. Let $F : C_0^{2,\alpha}(\overline{\Omega}) \to C^{0,\alpha}(\overline{\Omega})$ be a map defined by $\Delta u + f(u)$. Let u be a 1-transverse singularity which is not a fold. Then u is $(1,1)$-transverse if and only if

$$\begin{cases} \exists\, v \in T_u S \ : \\ \displaystyle\int_\Omega f'''(u)\, v\, \phi^3\, dx + 3 \int_\Omega f''(u)\, z\, \phi^2\, dx \neq 0 \\ \text{where } \phi \in \text{Ker } F'(u) \text{ and } z \in C_0^{2,\alpha}(\overline{\Omega}) \text{ is any} \\ \text{solution of the equation} \\ \Delta z + f'(u) z = - f''(u) \phi v \end{cases} \tag{33}$$

and u is a cusp singularity if and only if

$$\int_\Omega f'''(u) \phi^4\, dx + 3 \int_\Omega f''(u)\, z\, \phi^2\, dx \neq 0 \ , \tag{34}$$

$v = \phi$ and z as above.

For a proof of the proposition as well for one of the following normal form theorem we refer to Cafagna and Donati (1988b) (see also Berger, Church and Timourian (1985) and Lazzeri and Micheletti (1988)).

NORMAL FORM THEOREM FOR CUSPS. Let u be a cusp singularity of the map $F : X \to Y$. Then there exist a neighborhood U of u in X , a neighborhood V of $F(u)$ in Y , a Banach space Z and diffeomorphisms $\phi : U \to \mathbb{R}^2 \times Z$, $\psi : V \to \mathbb{R}^2 \times Z$ such that $(\psi \circ F \circ \phi^{-1})(t,s,z) = \tilde{F}(t,s,z) = (t^3+ts , s , z)$. That is to say, the following diagram commutes

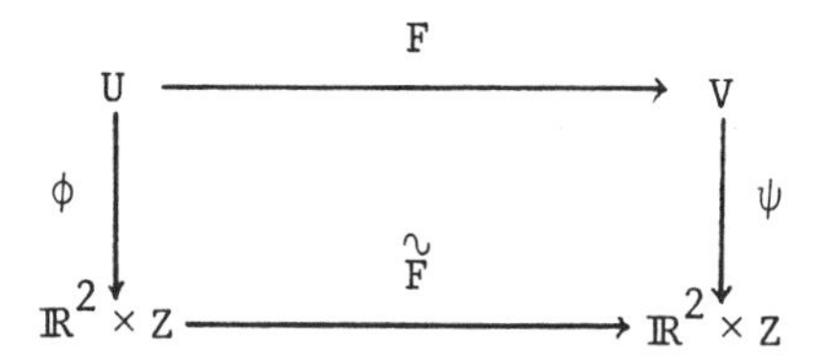

Many applications of cusp singularities to nonlinear differential equations have appeared in the last few years. Let us mention applications to nonlinear Dirichlet problems (Berger, Church and Timourian (1987), Lazzeri and Micheletti (1988), Cafagna and Tarantello (1987)), to minimal surfaces (Beeson and Tromba (1984)) to periodic solutions of first order nonlinear differential equations (Cafagna and Donati (1985)) and to Neumann problems (Ruf (1988)). In particular the last two papers contain global results, namely examples of differential problems which give rise to Fredholm maps which behave globally like the Whitney Cusp map $\mathbb{R}^2 \to \mathbb{R}^2$ defined by $(t,s) \to (t^3+ts , s)$ (Whitney (1955)).

In the following we will sketch an application of the analysis of singularities to the description of some qualitative features of a superlinear elliptic problem. Consider the problem

$$\begin{cases} \Delta u + u^3 = h & \text{in } \Omega \\ u = 0 & \text{on } \partial\Omega \end{cases} \tag{35}$$

where $\Omega \subset \mathbb{R}^n$ is a bounded domain with smooth boundary. It is well known that this problem has infinitely many solutions for every right-hand side if $n = 1$ (Ehrmann (1957), Fučik and Lovicar (1975). See also Bahri and Berestycki (1981)). Recently Bahri and Lions (1985 and 1987) were able to prove the same result in the case $n = 2$. It is a difficult open problem to decide whether this remains true in higher dimensions. The best known result is due to Bahri (1981) who proved that there are infinitely many solutions for generic (in some suitable sense) right-hand sides in dimension $n = 3$. All these papers made use of variational methods.

Let us consider now the map $C_o^{2,\alpha}(\overline{\Omega}) \to C^{o,\alpha}(\overline{\Omega})$ defined by $F(u) = \Delta u + u^3$, and try to figure out what does the singular set of F look like. As usual, one has

$$S = \left\{ u \in C_o^{2,\alpha}(\overline{\Omega}) \ : \ \exists\, v \in C_o^{2,\alpha}(\overline{\Omega}) \ : \ \Delta v + 3u^2 v = 0 \right\} \tag{36}$$

that is to say u is singular if and only if some eigenvalue of the eigenvalue problem (6) with weight $P = 3u^2$ is equal to 1. But this time the nonlinearity interacts with the whole spectrum of $-\Delta$, so that the singular set of F is the union of infinitely many S^j's (not all necessarily disjoint) defined by

$$S^j = \{ u \in C_o^{2,\alpha}(\overline{\Omega}) \ : \ 1 = \mu_j(3u^2) \} \tag{37}$$

A priori, the eigenvalues μ_j can be of arbitrarily large multiplicity and so the singular set will be a stratified manifold with strata of arbitrary codimensions. Nonetheless $S^1 = \{u : 1 = \mu_1(3u^2)\}$ is a nice set, i.e. a smooth hypersurface. The behaviour of the map F near S^1 can be fully described by simple singularity analysis, no matter how wildly F behaves elsewhere.

THEOREM 4. Let Ω be a bounded domain of $\mathbb{R}^n$ $(n \leq 3)$ with smooth boundary. Set $X = C_o^{2,\alpha}(\overline{\Omega})$ and $Y = C^{o,\alpha}(\overline{\Omega})$. Define S^1 as the set $\{u \in X : 1 = \mu_1(3u^2)\}$.

Then the map $F : X \to Y$ defined by $F(u) = \Delta u + u^3$ behaves in a neighborhood of S^1 like the Whitney Cusp map. To be more precise, define

$$S^1_\varepsilon = \{u \in X : |1-\mu_1(3u^2)| < \varepsilon\} \tag{38}$$

and consider the restricted map $F_\varepsilon : S_\varepsilon \to F(S_\varepsilon)$. Then, if ε is small enough, there exists a stratified hypersurface $Z = Z_1 \cup Z_2$ which disconnects $F(S_\varepsilon)$ in two components Y_1 and Y_3 such that

and

$$\begin{aligned} \sharp F_\varepsilon^{-1}(h) &= 1 \quad \text{if } h \in Y_1 \\ \sharp F_\varepsilon^{-1}(h) &= 3 \quad \text{if } h \in Y_3 \\ \sharp F_\varepsilon^{-1}(h) &= 1 \quad \text{if } h \in Z_1 \\ \sharp F_\varepsilon^{-1}(h) &= 2 \quad \text{if } h \in Z_2 \end{aligned} \tag{39}$$

If we moreover define $X_o = \{u \in X : 1 \leq \mu_1(3u^2)\}$ and $X_\varepsilon = X_o \cup S_\varepsilon$, then we have

$$\sharp F^{-1}(h) \cap X_\varepsilon \leq 3 \qquad \forall\, h \in F(X_\varepsilon) \tag{40}$$

Sketch of the proof. First of all observe that S^1 is a smooth hypersurface. In fact it is easy to see that the 1 -transversality condition

$$\exists\, v \in X : \int_\Omega u\, v\, \phi^2\, dx \neq 0 \tag{41}$$

is satisfied for every $u \in S^1$ (as usual ϕ is any nonzero function in Ker $F'(u)$). We note next that every singularity which is not a fold is actually a cusp. If we rewrite condition (32) in this case we obtain that $u \in S^1$ is a cusp if and only if

$$6 \int_\Omega \phi^4\, dx + 18 \int_\Omega u\, z\, \phi^2\, dx \neq 0 \tag{42}$$

where $z \in X$ is any solution of

$$\Delta z + 3u^2 z = -6u\phi^2 \tag{43}$$

Multiplying by z this last equation and integrating we obtain

$$\int_\Omega z\cdot\Delta z\, dx + 3\int_\Omega u^2 z^2\, dx = -6\int_\Omega u\, z\, \phi^2\, dx \quad . \qquad (44)$$

After integration by parts, substitution into the equation (42) yields the following form of the cusp condition

$$6\int_\Omega \phi^4\, dx + 3\left(\|z\|^2_{H^1_o} - \int_\Omega (3u^2)z^2\right) \neq 0 \quad . \qquad (45)$$

Now it is sufficient to observe that the second term of the left-hand side of (45) is positive by variational characterization of eigenvalues, to obtain that every singularity which is not a fold is a cusp.

In particular, remarking that at a cusp point the (1,1)-transversality condition is trivially satisfied, one has that the cusp locus, if non-empty, is a smooth hypersurface of S^1 . To sum up, once one has actually verified that the cusp locus is non-void, one has the following geometric picture of the first singular sheet : S^1 is a smooth hypersurface union of an open (in S^1) subset S^1_F , consisting of folds and a hypersurface (of S^1) S^1_C , consisting of cusps. To prove the theorem it will be enough to show that the map F is proper and that Z is the bijective image of S^1 , with $Z_1 = F(S^1_F)$ and $Z_2 = F(S^1_C)$. Then the multiplicity result will follow from the Normal Form Theorems.

We will not give a proof of the properness of F_ε . We only remark that this is a delicate matter. The properness can be deduced from a general result by Bahri and Lions (1988) , which essentially says that F is proper whenever restricted to the subsets $X_i = \{u \in X : 1 \leq \mu_{i+1}(3u^2)\}$. The proof of this is difficult. It seems very likely that in this case a much simpler proof of the properness could be given, based on geometric considerations.

The last step in the proof consists in showing that the map is 1-1 on S^1 . Actually we will prove more generally that F is 1-1 on the whole set $X_o = \{u \in X : 1 \leq \mu_1(3u^2)\}$. So this will also give a proof of formula (40).

Suppose that $F(u) = F(v)$ with u and v both in X_o. Then

$$\Delta(u-v) + (u^2+uv+v^2)(u-v) = 0 \tag{46}$$

i.e. 1 is some eigenvalue of the eigenvalue problem (6) with weight $P = u^2 + uv + v^2$ (say the j-th) :

$$1 = \mu_j(u^2 + uv + v^2) . \tag{47}$$

Now, remarking that $u^2 + uv + v^2 \leq \frac{3}{2}(u^2+v^2)$, using the Comparison Criteria and the concavity of μ_1, one obtains the following chain of inequalities

$$\begin{aligned} 1 &= \mu_j(u^2+uv+v^2) \geqq \mu_1(u^2+uv+v^2) \\ &\geq \mu_1\left(\frac{3}{2}(u^2+v^2)\right) > \frac{1}{2}\mu_1(3u^2) + \frac{1}{2}\mu_1(3v^2) \geq 1 . \end{aligned} \tag{48}$$

We note that the inequality before the last one is strict (thus yielding the desired contradiction) unless $u^2 = v^2$. But if $u = -v$, then $\Delta u + u^3 = 0$ and this implies $u = v = 0$, as it is easily seen.

3 FINITE SOLVABILITY

Let $F : X \to Y$ be a proper Fredholm map of index 0 and $h \in Y$ be a regular value of F. Then the set $F^{-1}(h)$ is finite (being discrete and compact). On the other hand nothing can be said *a priori* when h is a critical value. In this section we will discuss the problem of finiteness and give some examples of problems for which $F^{-1}(h)$ is actually finite for every $h \in Y$.

Consider the familiar semilinear Dirichlet problem

$$\begin{cases} \Delta u + f(u) = h & \text{in } \Omega \\ u = 0 & \text{on } \partial\Omega \end{cases} \tag{49}$$

In the last paragraph we gave examples of upper bounds for the cardinality of the set of solutions, found by singularity analysis techniques. In particular for the Ambrosetti and Prodi case (Thm. 3) this upper bound was shown to be 2. The main ingredients of the proof were : a) the convexity of f b) the fact that the interaction of the nonlinearity with the spectrum of $-\Delta$ was limited to the first eigenvalue.

It is easy to see that it is sufficient to violate the convexity condition in a small neighborhood of 0 to produce right-hand sides h such that $F^{-1}(h) \geqq 3$. For example one can prove that there is a constant $C > 0$, such that 0 is a cusp singularity of the map $F = \Delta + f : C_o^{2,\alpha}(\overline{\Omega}) \to C^{o,\alpha}(\overline{\Omega})$ provided that f satisfies the following assumptions

$$f(0) = 0 \ , \quad f'(0) = \lambda_1 \ , \quad f''(0) \neq 0 \ , \quad f'''(0) \neq f''(0)C \ . \tag{50}$$

In the one-dimensional case $\Omega = (0,\pi)$ the constant can be explicitly computed as $C = f''(0)\,\frac{5}{3}$ (Cafagna and Tarantello (1987)).

In general one expects to be able to produce (more or less) explicit conditions, involving the k-th order jet of f, for 0 to be a Morin singularity of codimension k (see Morin (1965) and Golubitsky and Guillemin (1973)). In this case there will exist right-hand sides h with $F^{-1}(h) \geqq k$ (Damon and Galligo (1976)). To sum up, provided that one has arbitrarily much patience in order to write down explicitly the conditions on f implying that 0 is a Morin singularity of an arbitrarily large codimension, then he can exhibit examples of very innocent-looking Dirichlet problems (interaction with the first eigenvalue only) with right-hand sides with arbitrarily large set of solutions.

Nevertheless if real analyticity is assumed the (possibly very large) sets of solutions stay always finite. Actually one can prove a more general result (interaction with any simple eigenvalue is allowed) :

THEOREM 5. Let λ_k be a single eigenvalue of the eigenvalue problem (6), where Ω is a bounded domain of $\mathbb{R}^n$ with smooth boundary and let $f : \mathbb{R} \to \mathbb{R}$ be a real-analytic function such that

$$\ell = \lim_{t \to \pm\infty} f'(t) \neq \lambda_k . \tag{51}$$

Suppose moreover that there exists a $\delta > 0$ such that

$$\lambda_{k-1} + \delta < f'(t) < \lambda_{k+1} - \delta \quad \forall t \in \mathbb{R} . \tag{52}$$

Then for every $h \in C^{o,\alpha}(\overline{\Omega})$ the set of solutions to the Dirichlet problem (49) is finite.

In order to give a proof of this theorem we need to develop some more tools of nonlinear functional analysis. In particular we will need the following version of the local representative theorem for real-analytic Fredholm map. Observe (Cafagna and Donati (1988a)) that a proof can be obtained combining the usual smooth Fredholm map version (Abraham and Robbin (1967)) with a real-analytic inverse function theorem (see Fucik et al. (1973)).

REAL-ANALYTIC LOCAL REPRESENTATIVE THEOREM. Let X and Y be Banach spaces, $F : X \to Y$ a real-analytic Fredholm map of index 0 and $u \in X$ a singularity of F with $\dim \operatorname{Ker} F'(u) = 1$. Then there exist a neighborhood U of u in X , V of $F(u)$ in Y , a Banach space Z and diffeomoprhisms $\phi : U \to \mathbb{R} \times Z$, $\psi : V \to \mathbb{R} \times Z$ such that

$$(\psi \circ F \circ \phi^{-1})(t,z) = \tilde{F}(t,z) = (\tilde{f}(t,z),z) \tag{53}$$

where $\tilde{f} : \mathbb{R} \times Z \to \mathbb{R}$ is a real-analytic function such that

$$\frac{\partial \tilde{f}}{\partial t}(0,0) = 0 . \tag{54}$$

We are going to use immediately the above normal form theorem to prove a simple but rather general result on the structure of the set of solutions to a real-analytic Fredholm equation with a singular right-hand side.

THEOREM 6. Let X and Y be Banach spaces and $F : X \to Y$ a real-analytic Fredholm map of index 0. Let h be a critical value of F and $C_1(h)$ be defined by

$$C_1(h) = \left\{ u \in S_F : \dim \operatorname{Ker} F'(u) = 1 \text{ and } u \text{ is an accumulation point of } F^{-1}(h) \right\} . \tag{55}$$

Then $C_1(h)$ is, if non-empty, a finite collection of real-analytic 1-manifolds.

Sketch of the proof. Let C be a component of $C_1(h)$ and $u \in C$. By the Real-Analytic Local Representative Theorem we know that F near u looks like a map $\tilde{F} : \mathbb{R} \times Z \to \mathbb{R} \times Z$ of the kind $\tilde{F}(t,z) = (\tilde{f}(t,z),z)$. Then $F^{-1}(h)$ is near u a real-analytic diffeomorphic image of the set $\tilde{F}^{-1}(0) = \{(t,0) \in \mathbb{R} \times Z : \tilde{f}(t,0) = 0\}$. Being u an accumulation point of $F^{-1}(h)$, it is easy to see that 0 is an accumulation point of the set of zeroes of the real-analytic function $\tilde{f}(\cdot,0) : \mathbb{R} \to \mathbb{R}$. Then $\tilde{f}(\cdot,0)$ is identically zero, so that $\tilde{F}^{-1}(0) = \mathbb{R} \times \{0\}$ and $F^{-1}(h)$ is locally a real-analytic diffeomorphic image of the real line. This is true for every $u \in C$ and so C is a real-analytic 1-manifold.

Observe that the tangent line to $C_1(h)$ at u is nothing but the kernel of the Fréchet derivative of F at u:

$$T_u C_1(h) = \operatorname{Ker} F'(u) . \tag{56}$$

If moreover $F : X \to Y$ is proper then $C_1(h)$ is also compact and so (by the classification of 1-manifolds (Milnor (1965))) $C_1(h)$ is a finite collection of circles.

We are almost ready to prove Theorem 5. The strategy will be to show that no circle can exist in the set of solutions. To this end the following Geometric Lemma will be very useful. We omit the simple proof.

GEOMETRIC LEMMA. Let X be a Banach space, $J : X \to \mathbb{R}$ a smooth functional and $C \subset X$ a smooth compact 1-manifold (i.e. a circle).

Then there exists a point $u \in C$, such that the tangent line to C at u is contained in the kernel of the gradient of J:

$$\exists\, u \in C \;:\; T_u C < \operatorname{Ker} J'(u) \;. \tag{57}$$

Sketch of the proof of Theorem 5. (See Cafagna and Donati (1988a) for the missing details). The hypotheses (51) and (52) imply, by means of the usual *a priori* estimates and Comparison Criteria that the map $F : C_o^{2,\alpha}(\bar{\Omega}) \to C^{o,\alpha}(\bar{\Omega})$ defined by $F = \Delta + f$ is a proper Fredholm map of index 0 such that for every singularity $u \in S_F$, $\dim \operatorname{Ker} F'(u) = 1$. Then, by Theorem 6, the set of accumulation points of $F^{-1}(h)$, if non-empty, is a finite collection of circles. To finish the proof it suffices to show that there are no circles in $F^{-1}(h)$. To this define a functional $J : C_o^{2,\alpha}(\bar{\Omega}) \to \mathbb{R}$ by

$$J(u) = \int_\Omega u\, \phi_k \tag{58}$$

where ϕ_k is an eigenfunction of $-\Delta$ relative to the eigenvalue λ_k. Suppose that there is a circle $C \subset F^{-1}(h)$. Then by the Geometric Lemma there exists an $u \in C$ such that

$$\operatorname{Ker} F'(u) = T_u C < \operatorname{Ker} J'(u) \tag{59}$$

that is to say, there exists a $v \in \operatorname{Ker} F'(u)$ such that

$$\int_\Omega v\, \phi_k = 0 \;. \tag{60}$$

But this is impossible. In fact it is not difficult to show (variational characterization of eigenvalues once more) that the only function in $\operatorname{Ker} F'(u)$ L_2-orthogonal to ϕ_k is 0.

To the author's knowledge the first general proof of finite solvability was contained in McKean and Scovel (1986). It concerned the two-point boundary value problem

$$\begin{cases} u'' + u^2 = k & \text{in } (0,\pi) \\ u(0) = u(\pi) = 0 \end{cases} \tag{61}$$

The approach by McKean and Scovel relied on the special geometry determined by the nonlinearity "u^2", while the machinery outlined above seems to have a certain flexibility. In fact the trick based on the combination of the Theorem 6 (to prove that non-isolated solutions form circles) and the Geometric Lemma (to prove that circles cannot exist) seems to be successful in different situations. It is just a matter of choosing with a grain of salt a functional which does the job (see Cafagna and Donati ((1986) and (1988a)) for other applications to semilinear elliptic problems, periodic solutions of first-order differential equations and orbits of plane conservative systems).

We finish the paper by resuming the superlinear elliptic problem considered at the end of section 2. This time we limit ourselves to consider the one-dimensional case

$$\begin{cases} u'' + u^3 = h & \text{in } (0,\pi) \\ u(0) = u(\pi) = 0 \end{cases} \tag{62}$$

As we already said, it is well known that there exist infinitely many solutions for every right-hand side. So the map $F : C_0^2([0,\pi]) \to C^0([0,\pi])$ defined by $F(u) = u'' + u^3$ is certainly not proper. Nevertheless the method described above can be combined with variational considerations to prove that there are only finitely many solutions with "fixed energy" and that there is only a discrete set of "possible energies" for each right-hand side h. To sum up one proves that for every $h \in C^0([0,1])$ the set of solutions is discrete. When this observation is combined with the known infinity results (Ehrmann (1957), Fučik and Lovicar (1975)) one has that the solutions form a divergent sequence.

THEOREM 7. Let $F : C_o^2([0,\pi]) \to C^o([0,\pi])$ be the map defined by $F(u) = u'' + u^3$. Then for every $h \in C^o([0,\pi])$, $F^{-1}(h)$ is a discrete set.

Sketch of the proof. Consider the functional $\Gamma : H_o^1([0,\pi]) \to \mathbb{R}$ defined by

$$\Gamma(u) = \int_o^{\pi} \left(\frac{1}{2}|u'|^2 - \frac{1}{4}u^4 - hu\right) dx \qquad (63)$$

As it is well known, u is a solution of (62) if and only if it is a critical point of Γ. We will show that the set $Cr(\Gamma)$ of critical points of Γ is discrete. Call $s \in \mathbb{R}$ a *critical level* if the level set

$$\Gamma_s = \{u \in H_o^1([0,\pi]) : \Gamma(u) = s\} \qquad (64)$$

contains a critical point of Γ. As we said before the proof will be in two steps : a) for every critical level s, the set $Cr(\Gamma) \cap \Gamma_s$ is finite b) critical levels are isolated.

We begin by observing that the functional Γ satisfies the condition of Palais and Smale

$$\text{P-S} \quad \begin{cases} \text{Every sequence } \{u_n\} \text{ such that } \Gamma(u_n) \text{ is bounded} \\ \text{and } \Gamma'(u_n) \to 0 \text{ admits a converging subsequence} \end{cases} . \qquad (65)$$

Proof of a). The P-S condition in particular implies that the set $Cr(\Gamma) \cap \Gamma_s$ is compact for every s. Then a trivial adjustment of the argument of Theorem 6 proves that the set of non-isolated critical points at a given level s is, if non-empty, a finite set of circles. To show that this is actually impossible we put into play a new functional $J : C_o^2([0,\pi]) \to \mathbb{R}$ defined by

$$J(u) = u'(0) . \qquad (66)$$

Then the Geometric Lemma implies the existence on every circle C of a u such that

$$\text{Ker } F'(u) = T_uC < \text{Ker } J'(u) \quad . \tag{67}$$

Let $v \in \text{Ker } F'(u)$ be such that $\|v\|_{L^2} = 1$, then formula (67) means that for every $t \in \mathbb{R}$, the function tv is a solution of the Cauchy problem

$$\begin{cases} tv'' + 3u^2 tv = 0 \\ tv(0) = 0 \\ tv'(0) = 0 \end{cases} \tag{68}$$

and this is obviously absurd.

Proof of b). Suppose that s is an accumulation point of critical levels, namely that there exists a sequence $\{u_n\} \subset H_0^1([0,\pi])$ such that

$$\begin{aligned} \Gamma'(u_n) &= 0 \\ \Gamma(u_n) &\to s \end{aligned} \tag{69}$$

Then the P-S condition implies the existence of a subsequence converging to an u_o with $\Gamma(u_o) = s$. So u_o is an accumulation point of critical points of Γ . Then by Theorem 6 the critical set of Γ is in a neighborhood of u_o a real-analytic curve C . But $\Gamma'(u) \equiv 0$ on C , so that Γ has to be constant on C and this contradicts the fact that s is an accumulation point of critical levels.

REFERENCES

Abraham, R. and Robbin, J. (1967). Transversal Mappings and Flows, Benjamin, New York - Amsterdam.

Ambrosetti, A. and Prodi, G. (1972). On the inversion of some differentiable mappings with singularities between Banach spaces, Ann. Mat. Pura Appl., 93 : 231-246.

Bahri, A. (1981). Topological results on a certain class of functionals and application, J. Funct. Anal., 41 : 397-427.

Bahri, A. and Berestycki, H. (1981). A perturbation method in critical point theory and applications, Trans. Amer. Math. Soc., 267, 1 : 1-32.

Bahri, A. and Lions, P.L. (1985). Remarques sur la théorie variationnelle des points critiques et applications, C.R. Acad. Sci. Paris, 301 : 145-147.

Bahri, A. and Lions, P.L. (1987). Morse index of some min-max critical points. I. Applications to multiplicity results, preprint.

Bahri, A. and Lions, P.L. (1988). In preparation.

Banach, S. and Mazur, S. (1934). Ueber mehrdeutige stetige Abbildungen. Studia Math., 5 : 174-178.

Beeson, M.J. and Tromba, A.J. (1984). The cusp catastrophe of Thom in the bifurcation of minimal surfaces. Manuscripta Math., 46 : 273-308.

Berger, M.S. (1977). Nonlinearity and Functional Analysis, Academic Press, New York - San Francisco - London.

Berger, M.S. and Church, P.T. (1979). Complete integrability and perturbation of a nonlinear Dirichlet problem I, Indiana Univ. Math. J., 28 : 935-952. Erratum (1981). Ibid 30 : 799.

Berger, M.S., Church, P.T. and Timourian, J.G. (1985). Folds and cusps in Banach spaces, with applications to nonlinear partial differential equations, I, Indiana Univ. Math. J., 34 : 1-19.

Berger, M.S., Church, P.T. and Timourian, J.G. (1987). Folds and cusps in Banach spaces, with applications to nonlinear partial differential equations, II, preprint.

Borisovich, Y., Zvyagin, V. and Sapronov, Y. (1977). Nonlinear Fredholm maps and the Leray-Schauder theory, Russ. Math. Surveys, 32, 4 : 3 -54.

Caccioppoli, R. (1932). Un principio di inversione per le corrispondenze funzionali e sue applicazioni alle equazioni a derivate parziali, Rend. Acc. Naz. Lincei, VI-16 : 390-395, 484-489.

Caccioppoli, R. (1936). Sulle corrispondenze funzionali inverse diramate : teoria generale e applicazioni ad alcune equazioni funzionali nonlineari e al problema di Plateau, Rend. Acc. Naz. Lincei, VI-24 : 258-263, 416-421.

Cafagna, V. (1984). On fold singularities of Fredholm maps and a theorem by Ambrosetti and Prodi, Convegno di Topologia, L'Aquila, Suppl. Rend. Circolo Mat. Palermo, II-4 : 21-30.

Cafagna, V. (1986). Oscillations forcées, preprint.

Cafagna, V. and Donati, F. (1985). Un résultat global de multiplicité pour un problème différentiel non linéaire du premier ordre, C.R. Acad. Sci. Paris, 300-I-15 : 523-526.

Cafagna, V. and Donati, F. (1986). Finitude du nombre des solutions de quelques problèmes aux limites non linéaires, C.R. Acad. Sci. Paris, 303-I-17 : 857-860.

Cafagna, V. and Donati, F. (1988a). On the finite solvability of some nonlinear differential problems, preprint.

Cafagna, V. and Donati, F. (1988b). Singularity theory and the number of solutions to some simple nonlinear first order equations, preprint.

Cafagna, V. and Tarantello, G. (1987). Multiple solutions for some semilinear elliptic equations, Math. Ann. 276 : 643-656.

Courant, R. and Hilbert, D. (1953). Methods of mathematical physics, Vol. I, Wiley, New York.

Damon, J. (1986). A theorem of Mather and the local structure of nonlinear Fredholm maps, Proc. Symp. Pure Math., 45-1 : 339-352.

Damon, J. (1987). Time dependent nonlinear oscillations with many periodic solutions, Siam J. Math. Anal., 18-5 : 1294-1316.

Damon, J. and Galligo, A. (1976). A topological invariant for stable map germs, Inv. Math. 32 : 103-132.

Dolph, C.L. (1949). Nonlinear integral equations of the Hammerstein type, Trans. Am. Math. Soc., 60 : 289-307.

Ehrmann, H. (1957). Über die Existenz der Lösungen von Randwertaufgaben bei gewönlicher nichtlinearen Differentialgleichungen zweiter ordnung, Math. Ann., 134 : 167-194.

Fučik, S. and Lovicar, V. (1975). Periodic solutions of the equation $x''(t) + g(x(t)) = p(t)$, Časopis Pěst. Mat., 100 : 160-175.

Fučik, S., Nečas, J., Souček, J. and Souček, V. (1973). Spectral Analysis of Nonlinear Operators, Springer-Verlag, Berlin Heidelberg New York.

Greenberg, M.J. (1966). Lectures on Algebraic Topology, Benjamin, Meulo Park, Ca.

Golubitsky, M. and Guillemin, V. (1973). Stable Mappings and Their Singularities, Springer-Verlag, New York, Heidelberg, Berlin.

Hammerstein, A. (1929). Nichtlineare Integralgleichungen nebst Anwendungen, Acta Math., 54 : 117-176.

Lazzeri, F. and Micheletti, A.M. (1988). An application of singularity theory to nonlinear differentiable mappings between Banach spaces, Nonlinear Analysis, T.M.A., to appear.

Martinet, J. (1982). Singularities of Smooth Functions and Maps, Cambridge University Press, Cambridge.

McKean, H.P. (1987). Singularities of a simple elliptic operator, J. Diff. Geom. 25 : 157-165.

McKean, H.P. and Scovel, J.C. (1986). Geometry of some simple nonlinear differential operators, Ann. Scuola Norm. Sup. Pisa, IV-XIII-2 : 300-346.

Milnor, J.W. (1965). Topology from the Differentiable Viewpoint, The University Press of Virginia, Charlottesville.

Morin, B. (1965). Formes canoniques des singularités d'une application différentiable, C.R. Acad. Sci. Paris, 260-1 : 5662-5665, 6503-6506.

Plastock, R. (1974). Homeomorphisms between Banach spaces, Trans. Am. Math. Soc. 200 : 169-183.

Prodi, G. and Ambrosetti, A. (1973). Analisi non Lineare, Editrice Tecnico Scientifica, Pisa.

Ruf, B. (1988). In preparation.

Vainberg, M.M. (1968). Metodo variazionale e metodo di Caccioppoli nella teoria delle equazioni funzionali non lineari, Symp. Math., II, Academic Press, London - New York, pp. 211-226.

Whitney, H. (1955). On singularities of mappings of Euclidean spaces I, mappings of the plane into the plane, Ann. of Math., 62 : 374-410.

Boundary Conditions and Vanishing Index for α-Contractions and Condensing Maps

G. Fournier
Departement de Mathématique
Université de Sherbrooke
Sherbrooke, Canada

M. Martelli
Mathematics Department
California State University
Fullerton, CA

1. Introduction

Let Ω be a bounded open neighborhood of the origin in a Banach space E and let $f : \overline{\Omega} \to E$ be continuous and condensing. The main purpose of this paper is to investigate what additional conditions on f will ensure that $\text{ind}(E, f, \Omega) = 0$ and to derive from this fact some useful information on the existence of positive eigenvalues and corresponding eigenvectors for f, belonging to the boundary of Ω. We also examine the case where Ω is a bounded open neighborhood of the origin on a cone K or on a wedge W of E, and we provide an affirmative answer to a conjecture raised about ten years ago by I. Massabo-C. Stuart [10]. Our work culminates a series of efforts, which were only partially successful (see [1], [9], [12]), aimed at obtaining a condition on the boundary behavior of f, which ensures that $\text{ind}(E, f, \Omega) = 0$ or, more generally $\text{ind}(K, f, \Omega) = 0$. The key element of our approach is the so-called mod p theorem of M.A. Krasnoselskii-P.P. Zabreiko [7] and H. Steinlein [17] (see also [4], [15] and [16]).

2. Notations and Definitions

A *wedge* W of a Banach space E is a closed subset of E such that

$$\mathbf{x}, \mathbf{y} \in W, \quad a, b \geq 0 \text{ implies } ax + by \in W.$$

A *cone* K of E is a wedge which satisfies the additional condition

$$\mathbf{x} \in K \text{ and } -\mathbf{x} \in K \text{ implies } \mathbf{x} = \mathbf{0}.$$

A subset Ω of W is *relatively open* if there exists an open set $V \subset E$ such that

$$\Omega = V \cap W.$$

The *relative boundary* of Ω, denoted by $\partial\Omega$, is the set

$$\partial\Omega = \overline{\Omega} \cap \overline{\Omega}_c \cap W$$

where $\overline{\Omega}$ is the closure of Ω and $\Omega_c = E \backslash \Omega$. If no confusion arises we shall simply say that Ω is open in W and $\partial\Omega$ is its boundary.

Let $\mathcal{B}$ be the family of all bounded subsets of E. A function $\theta : \mathcal{B} \to [0, +\infty)$ is said to be a *generalized measure* of non-compactness if

(1) $\theta(A) = 0$ if and only if $\overline{A}$ is compact;

(2) $\theta(\overline{co}\ A) = \theta(A)$, where $\overline{co}\ A$ denotes the convex closure of A;

(3) $\theta(A \cup B) = \max\{\theta(A), \theta(B)\}$;

(4) $\theta(tA) = t\theta(A)$ for every $t \geq 0$;

(5) $\theta(A + B) \leq \theta(A) + \theta(B)$.

The function $\alpha : \mathcal{B} \to [0, +\infty)$ defined by $\alpha(A) = \inf\{\varepsilon > 0$: A admits a finite covering of sets whose diameter does not exceed $\varepsilon\}$, was first proposed by C. Kuratowski [8] and it satisfies (1)-(5) (see G. Darbo [2] for the first proof of (2)). If the word "sets" is replaced by "spheres" we obtain a different measure of non-compactness, which is usually denoted by γ. It is easily seen that

$$\gamma(A) \leq \alpha(A) \leq 2\gamma(A).$$

In [6] M. Furi-A. Vignoli proved that the unit sphere S of E has the property $\alpha(S) = 2$.

In this paper we shall use the Kuratowski measure of non-compactness. The results are valid for any generalized measure of non-compactness.

A function $f : E \to E$ is called *condensing* if $A \in \mathcal{B}$ and $\alpha(A) > 0$ implies

$$\alpha(f(A)) < \alpha(A)$$

We shall say that f is an α-contraction if the stronger inequality

$$\alpha(f(A)) \leq p\alpha(A) \qquad p < 1$$

holds; and that f is α-Lipschitz with constant p if the above inequality holds only with $p \geq 1$. A function $f : E \to E$ may be condensing or α-contractive one a cone K of E without being so in the entire space E. Examples of such functions can be easily constructed, but they are also known to arise in many applications.

3. Results

Our first theorem is stated for compact maps. The result is not new and it is included here for completeness and for comparison with successive results regarding noncompact maps.

Theorem 3.1. *Let Ω be an open bounded neighborhood of the origin in a Banach space E and let $f : \overline{\Omega} \to E$ be continuous and compact. Assume that*

$$f(\partial\Omega) \cap co\overline{\Omega} = \emptyset. \tag{3.1}$$

Then

$$\text{ind}(E, f, \Omega) = 0. \tag{3.2}$$

Proof. Let $\varepsilon > 0$ be such that the ball $B(0, \varepsilon) = \{x \in E : ||x|| \leq \varepsilon\}$ is contained in Ω. It is known that there exists a continuous map $r : E \to E$ such that $||r(x)|| = \varepsilon$ for every $||x|| < \varepsilon$ and $r(x) = x$ for every $||x|| \geq \varepsilon$. The map $g(x) = r(f(x))$ is compact, coincides with f on $\partial\Omega$ and is such that $g(x) \neq 0$ for all $x \in \overline{\Omega}$. Let $\rho = \sup\{||x|| : x \in \partial\Omega\}$ and define $h(x) = \frac{2\rho}{\varepsilon} g(x)$. Then h is compact, homotopic to g by a homotopy without fixed points on $\partial\Omega$ and such that $h(\overline{\Omega}) \cap \overline{\Omega} = \emptyset$. Thus

$$\text{ind}(E, f, \Omega) = \text{ind}(E, h, \Omega) = 0.$$

Remark 3.1. The conditions

1. there exists $\delta > 0$ such that $||f(x)|| \geq \delta$ for every $x \in \partial\Omega$; and
2. $tf(x) \neq x$ for all $x \in \partial\Omega$ and $t \geq 1$;

are sometimes found in the literature in place of (3.1). The two theorems with the sets of conditions are equivalent. In fact it is obvious that (3.1) implies 1. and 2., which in turns imply (3.1) via the homotopy $\left[1+\left(\frac{2\rho}{\delta}-1\right)t\right]f(x)$. Hence

Theorem 3.2. *Let Ω be a bounded open neighborhood of the origin in a Banach space E and let $f:\overline{\Omega}\to E$ be continuous and compact. Assume that*

$$\inf\{||f(x)||: x\in\partial\Omega\}=\delta>0 \tag{3.3}$$

$$tf(x)\neq x \quad \text{for all} \quad x\in\partial\Omega,\quad t\in[0,1]. \tag{3.4}$$

Then

$$\operatorname{ind}(E,f,\Omega)=0. \tag{3.5}$$

We shall prove now that the same condition (3.1) is sufficient to establish (3.2) where f is an α-contraction with constant $p<1$. The proof is now different since the composition rf need not be an α-contraction any longer.

Theorem 3.3. *Let Ω be a bounded open neighborhood of 0 in E and $f:\overline{\Omega}\to E$ be α-contractive with constant $p<1$. Assume that*

$$f(\partial\Omega)\cap \mathrm{co}\overline{\Omega}=\emptyset. \tag{3.6}$$

Then

$$\operatorname{ind}(E,f,\Omega)=0. \tag{3.7}$$

Proof. Let $q>1$ be such that $qp<1$. Then there exists $\varepsilon>0$ such that $||x-y||>2\varepsilon$ for every $x\in co\overline{\Omega}$ and $y\in qf(\partial\Omega)=g(\partial\Omega)$. In fact define the Minkovski functional [12] $\rho:E\to[0,+\infty)$ by

$$\rho(x)=\inf\{t:\frac{x}{t}\in\overline{co\Omega}\}.$$

Then ρ is a norm in E and for every $y \in qf(\partial\Omega)$ we have

$$||y - x||_\rho \geq q - 1 > 0.$$

Since the ρ-norm is equivalent to the original norm there exists $m > 0$ such that

$$||y - x|| \geq m||y - x||_\rho.$$

Hence we can choose $2\varepsilon < m(q - 1)$.

Now select m so large that

$$p^m \alpha(\Omega) < \varepsilon$$

and define

$$V_m = g^{1-m}(\Omega).$$

Notice that by (3.6), $g^{-1}(\Omega) \subset \Omega$, and by induction $V_m \subset V_{m-1} \subset \cdots \subset V_1 = \Omega$. Moreover

$$g^m(\partial V_m) \subset g(\partial\Omega).$$

In fact, by elementary topology, we have for a continuous map h,

$$h(\partial h^{-1}U) \subset \partial U.$$

Thus with $h = g$ and $U = g^{1-m}(\Omega)$ we obtain

$$g(\partial g^{-m}\Omega) \subset \partial g^{1-m}(\Omega) \quad m \geq 1.$$

By applying g^m to both sides we get

$$g^{m+1}(\partial V_{m+1}) \subset g^m(\partial V_m).$$

Therefore

$$g(\partial\Omega) \supset g^2(\partial V_2) \supset g^3(\partial V_3) \cdots \supset g^m(\partial V_m).$$

Since $\alpha g^m(V_m) \leq p^m \alpha(\Omega) < \varepsilon$, we can cover $g^m(V_m)$ with finitely many closed convex sets $K_1, \cdots, K_s$ such that diam $K_i < \varepsilon$, $i = 1, 2, \cdots, s$ and we can construct (see [5]) a continuous and compact map $\pi : \cup_i K_i \to \cup_i K_i$ such that $\pi(K_i) \subset K_i$.

Then πg^m and g^m are homotopic. By Theorem 1 we have $\text{ind}(E, \pi g^m, V_m) = 0$. Thus $\text{ind}(E, g^m, V_m) = 0$.

Without loss of generality we can assume that m is a prime number and it is larger then $|\text{ind}(E, g, \Omega)|$. From the mod p theorem ([5] and [13]) we obtain

$$\text{ind}(E, g, \Omega) \equiv \text{ind}(E, g^m, V_m) \mod m.$$

Thus $\text{ind}(E, g, \Omega) = nm$ for some integer n. Since $|\text{ind}(E, g, \Omega)| < m$, the only possibility is $n = 0$. The map g and f are obviously homotopic without fixed points on $\partial\Omega$ and the result follows.

Remark 3.2. In the proof of Theorem 3.3 the boundaries are taken in E, $\overline{\Omega}$, $g^{-1}(\overline{\Omega})$, etc. However, since $g^{1-m}(\overline{\Omega})$ is a closed subset of E and V_m is an open subset of $g^{1-m}(\overline{\Omega})$, it follows that the boundary of V_m in E coincides with its boundary in $g^{1-m}(\overline{\Omega})$.

Remark 3.3. In Theorems 3.1 and 3.3 we can replace the condition "$f(\partial\Omega) \cap co\overline{\Omega} = \emptyset$" with the more general assumption

$$(3.6') \qquad f(\partial\Omega) \cap co\Omega = \emptyset \text{ and } x \neq f(x) \text{ for all } x \in \partial\Omega \cap \partial co\Omega.$$

The proofs are exactly the same.

In the compact case we have seen that the condition $f(\partial\Omega) \cap co\overline{\Omega} = \emptyset$ is equivalent to (see Remark 3.1)

1. $\inf\{||f(x)|| : x \in \partial\Omega\} = \delta > 0$;
2. $tf(x) \neq x$ for all $x \in \partial\Omega$, $t \geq 1$.

In the non-compact case condition 1 needs some adjustment, as the following example shows.

Example 1. Let $E = \ell^2$, $f : E \to E$; $x \to \frac{1}{2}(0, x_1, x_2, \cdots)$ and $\Omega = \{x \in \ell^2 : ||x|| < 1\}$. Then f is an α-contraction with constant $k = 1/2$. Conditions 1 and 2 are obviously satisfied, but $\text{ind}(E, f, \Omega) = 1$, since f is homotopic to the constant map which sends everything into 0.

We are now interested in seeing which adjustments can be made in the non-compact case.

Let $\rho = \sup\{||x|| : x \in \partial\Omega\}$ and let $\rho p < \delta$, where p is the α-contraction constant of f. Define $g : \overline{\Omega} \to E$ by $g(x) = \frac{f(x)}{p+\varepsilon}$, where $\varepsilon > 0$ is such that $p + \varepsilon < 1$

and $\rho(p+\varepsilon) < \delta$. Then g is an α-contraction with constant $\frac{p}{p+\varepsilon}$. Moreover

$$\delta' = \inf\{||g(x)|| : x \in \partial\Omega\} = \frac{\delta}{p+\varepsilon} > \rho.$$

Thus $g(\partial\Omega) \cap co\overline{\Omega} = \emptyset$. If we assume that $tf(x) \neq x$ for all $x \in \partial\Omega$, $t \geq 1$ then f and g are homotopic via the admissible homotopy $H : \overline{\Omega} \times [0,1] \to E$ defined by

$$H(x,r) = (1-r)f(x) + rg(x).$$

Thus

$$\operatorname{ind}(E, f, \Omega) = \operatorname{ind}(E, g, \Omega).$$

Since g satisfies the assumption of Theorem 3.2 we have the following result.

Theorem 3.4. *Let Ω be an open bounded neighborhood of 0 in E and let $f : \overline{\Omega} \to E$ be an α-contraction with constant $p < 1$. Let*

$$\delta = \inf\{||f(x)|| : x \in \partial\Omega\}, \quad \rho = \sup\{||x|| : x \in \partial\Omega\}.$$

Assume that

$$\delta > \rho p \tag{3.8}$$

$$tf(x) \neq x \quad \text{for all} \quad x \in \partial\Omega, \quad t \geq 1. \tag{3.9}$$

Then

$$\operatorname{ind}(E, f, \Omega) = 0. \tag{3.10}$$

Notice that for $p = 0$, Theorem 3.4 reduces to Theorem 3.2. Theorems 3.1-3.4 can be extended in a straightforward manner to the case when Ω is an open, bounded neighborhood of the origin in a cone $K \subset E$. We only need to assume that f maps $\overline{\Omega}$ into K, and to keep in mind that $\overline{\Omega}$, $\partial\Omega$ are to be considered relative to the cone. The index is well defined (see, for example R.D. Nussbaum [12]) and only minor changes are required in the proofs. For example the following version of Theorem 3.4 holds

Theorem 3.5. *Let $K \subset E$ be a cone and Ω be an open, bounded neighborhood of* 0 *in K Let $f : \overline{\Omega} \to K$ be an α-contraction with constant p and*

$$\delta = \inf\{||f(x)|| : x \in \partial\Omega\}; \rho = \sup\{||x|| : x \in \partial\Omega\}$$

where $\overline{\Omega}$ and $\partial\Omega$ are taken with respect to K. Assume that

$$\delta > p\rho \tag{3.11}$$

$$tf(x) \neq x \quad \text{for all} \quad x \in \partial\Omega, \quad t \geq 1. \tag{3.12}$$

Then

$$\text{ind}(K, f, \Omega) = 0.$$

In [10] I. Massabo-Ch. Stuart proved that an α-contraction $f : \overline{\Omega} \to K$, where Ω is an open neighborhood of the origin in a normal cone K contained in a Banach space E has an eigenvector $x_0 \in \partial\Omega$, corresponding to a positive eigenvalue λ, provided that

$$\delta\gamma > p\rho \tag{3.11}$$

where δ, ρ, p have the same meaning as in Theorem 3.4, and γ is the normality constant of the cone, i.e., a constant such that

$$||x + y|| \geq \gamma ||x||$$

for all $x, y \in K$. Notice that $\gamma \leq 1$. They conjecture that γ was not needed in (3.11). R.D. Nussbaum [12] proved that their conjecture was correct at least when the open set Ω is star-shaped with respect to the origin, i.e., $tx \in \Omega$ for every $x \in \Omega$ and $t \in [0, 1]$. M. Martelli [9] proved that their result was true for every cone K, by replacing the normality constant with the quasinormality constant (see also W.V. Petryshyn [13]), i.e., a positive number γ such that

$$||x + \lambda x_0|| \geq \gamma ||x||$$

for some $x_0 \in K$ and all $x \in K$, $\lambda \geq 0$.

As a consequence of Theorem 3.5 we see that the conjecture of Massabo-Stuart does not need any further assumption. It is true in its full generality. More precisely we have

Corollary 3.1. *Let Ω be a bounded, open neighborhood of the origin in a cone K of a Banach space E. Let $f : \overline{\Omega} \to K$ be an α-contraction with constant p such that*

$$\delta > p\rho$$

with δ, ρ as in Theorem 3.5. Then there exists $x_0 \in \partial\Omega$ and $\lambda \geq \frac{\delta}{\rho}$ such that

$$f(x_0) = \lambda x_0.$$

Proof. Assume that $f(x) \neq \lambda x$ for all $x \in \partial\Omega$ and $\lambda \geq \frac{\delta}{\rho}$. If $\lambda < \frac{\delta}{\rho}$ then

$$||\lambda x|| < \frac{\delta}{\rho}||x|| \leq \delta \leq ||f(x)||.$$

Thus $f(x) \neq \lambda x$ for all $\lambda \geq 0$ and $x \in \partial\Omega$. Then, by Theorem 3.5

$$\text{ind}(K, f, \Omega) = 0.$$

On the other hand f is homotopic to the constant map $h(x) = 0$ for all $x \in \Omega$. Thus

$$\text{ind}(K, f, \Omega) = 1.$$

This contradiction shows that $f(x) = \lambda x$ for some $x_0 \in \partial\Omega$ and $\lambda \geq \frac{\delta}{\rho}$.

When Ω is a ball centered at the origin, or the intersection of such a ball with the cone K, the proofs of Theorems 3.3-3.5 can be done in a different manner without using the mod-p theorem. Since these proofs are simpler and more direct we present here the one of Theorem 3.4.

Proof of Theorem 3.4 when $\Omega = \{x \in E : ||x < r\}$. Without loss of generality we may assume that $||f(x)|| \geq r_1 > r$ for every $||x|| = r$, and that there exists r_2 with the property

$$2||f(x)|| < r_2$$

for all $||x|| = r$. Let $\pi : D(0, r_2) \to E$, be defined by

$$\pi(x) = \begin{cases} x & \text{if } ||x|| \leq r \\ r\frac{x}{||x||} & \text{if } ||x|| \geq r. \end{cases}$$

The map $f \circ \pi$ is an α-contraction and it is homotopic to a constant via the admissible homotopy $H(x, t) = tf(\pi(x))$.

There are no fixed points of $f \circ \pi$ in $S_r = \{x \in E : ||x|| = r\}$. Moreover $f \circ \pi$ maps the annulus $A = \{x : r < ||x|| \leq r_2\}$ into itself. Since A is contractible we have

$$\text{ind}(D(0, r_2), f \circ \pi, A) = \Lambda(f \circ \pi, A) = 1,$$

where
$\Lambda(f \circ \pi, A)$ is the Lefschetz number of $f \circ \pi$ relative to A. Similarly $\text{ind}(D(0, r_2),$ $f \circ \pi, D(0, r_2)) = 1$. Hence, by additivity,

$$\text{ind}(D(0, r_2), f, B(0, r)) = \text{ind}(D(0, r_2), f \circ \pi, B(0, r)) = 0.$$

We now turn to the case when $f : \overline{\Omega} \to E$ (or $f : \overline{\Omega} \to K$) is condensing, i.e.,

$$\alpha(f(A)) < \alpha(A)$$

for every $A \subset \overline{\Omega}$. The proof of Theorem 3.3 need to be modified since we do not have a constant $p < 1$ to use for defining g. This difficulty is bypassed with the following, a bit more time consuming, approach.

Define

$$V_m = f^{1-m}(\Omega)$$

and notice that

$$\alpha(f^m(\partial V_m)) \leq \alpha f^m(\overline{V}_m) \leq \alpha f^m(\overline{\Omega}).$$

Moreover, as we have seen previously in the proof of Theorem 3.3 we have the inclusion

$$f^{m+1}(\partial V_{m+1}) \subset f^m(\partial V_m)$$

for $m \geq 1$.

It is also known that whenever f is a condensing map and A is a bounded set, then $\alpha(f^n(A)) \to 0$ as $n \to +\infty$ (see P. Massat [11]). Therefore, by a Theorem of Kuratowski [8] we obtain that

$$K = \cap_m \overline{f^m(\partial V_m)}$$

is nonempty, compact and obviously contained in $\overline{f(\partial\Omega)}$. Therefore we have proved the following result.

Lemma 3.1. *Let Ω be a bounded open set and $f : \overline{\Omega} \to E$ be condensing. Define*

$$V_m = f^{1-m}(\Omega) \qquad m \geq 1.$$

Then

$$K = \cap_m \overline{f^m(\partial V_m)}$$

is nonempty, compact and contained in $\overline{f(\partial\Omega)}$.

We now assume that $f(\partial\Omega) \cap (co\ \Omega \cup \partial\Omega) = \emptyset$ (this is implied by (3.6') but implies (3.6)).

Clearly if $x \in \partial\Omega$ then $f(x) \notin \overline{\Omega}$ that is $x \notin f^{-1}(\overline{\Omega})$ and so we get that $x \notin \overline{V}_2$. Thus $\overline{V}_2 \subset \Omega$.

Since $K \subset \overline{f(\partial\Omega)} \subset E \backslash \Omega$ is compact, we have $K \cap \overline{V}_2 = \emptyset$ and $3d = d(K, \overline{V}_2) > 0$.

Define $M = 1 + \sup\{||x|| : x \in K\}$ and choose $1 < q < 1 + \frac{d}{M}$. Since $qK \subset E \backslash \overline{co}(\Omega)$, as can be seen from the first part of the proof of Theorem 3.3, there exists $0 \lneq \varepsilon < \max\{\frac{1}{3}, \frac{d}{2}\}$ such that $qN_{2\varepsilon}(K) \subset E \backslash \overline{co}(\Omega)$ where $N_h(K) = \{x \in E : d(x, K) \leq h, h > 0\}$.

Note that from Lemma 3.1, there exists n large enough such that $\overline{f^m(\partial V_m)} \subset N_\varepsilon(K)$ and $\alpha(f^m(\overline{V}_m)) < \varepsilon$ for all $m > n$.

We can now cover $f^m(\overline{V}_m))$ by closed convex sets of diameter less than ε to obtain, as in the proof of Theorem 3.3, a compact map $\pi_m f^m$ such that $||f^m(x) - \pi_m f^m(x)|| < \varepsilon$ for all $x \in \overline{V}_m$.

Moreover, since $\pi f^m(\partial V_m) \subset N_{2\varepsilon}(K)$, the compact map $q\pi_m f^m$ satisfies $q\pi_m f^m(\partial V_m) \cap \overline{co}(\Omega) = \emptyset$.

Thus $q\pi_m f^m(\partial V_m) \cap \overline{co}(V_m) = \emptyset$, which implies by Theorem 3.1, that $\text{ind}(E, q\pi_m f^m, V_m) = 0$.

Let us prove that the index above is equal to the index of f^m on V_m.
In fact, since for all $t \in [0,1]$ and $x \in \partial V_m$

$$h(x,t) = (1-t)f^m(x) + t\pi_m f^m(x) \in N_{2\varepsilon}(K) \subset N_d(K) \subset E\backslash\overline{V}_2 \subset E\backslash\overline{V}_m,$$

h is an admissible homotopy and

$$\text{ind}(E, \pi_m f^m, V_m) = \text{ind}(E, f^m, V_m).$$

Similarly the homotopy $H(x,t) = [1 + t(q-1)]\pi_m f^m(x) \in N_{2\varepsilon+d}(K) \subset N_{2d}(K) \subset E\backslash\overline{V}_2 \subset E\backslash\overline{V}_m$, $t \in [0,1]$ and $x \in \partial V_m$ is admissible since $||t(q-1)\pi_m f^m(x)|| \leq (q-1)[2\varepsilon + \sup\{||f^m(x)|| : x \in \partial V_m\}] \leq (q-1)[2\varepsilon + \varepsilon + \sup\{||x|| : x \in K\}] = (q-1)M < d$. Hence

$$\text{ind}(E, q\pi_m f^m, V_m) = \text{ind}(E, \pi_m f^m, V_m).$$

Finally, as at the end of the proof of Theorem 3.3, we obtain $\text{ind}(E, f, \Omega) = 0$. We have proved the following result.

Theorem 3.6. *Let Ω be a bounded open neighborhood of the origin in a Banach space E and let $f : \overline{\Omega} \to E$ be condensing. Assume that*

$$f(\partial\Omega) \cap (co\Omega \cup \partial\Omega) = \emptyset.$$

Then

$$\text{ind}(E, f, \Omega) = 0.$$

♦

The above result holds when Ω is contained in a cone K and $f : \overline{\Omega} \to K$.

4. The index for the multivalued case

A multivalued map $F : \overline{U} \to E$, is upper semicontinuous if $F(x)$ is compact for all $x \in \overline{U}$ and $F^{-1}(W) = \{x : F(x) \subset W\}$ is open in $\overline{U}$ for all W open in E. F is said to be acyclic if for all $x \in \overline{U}$, $F(x)$ is acyclic with respect to the Cech homology. This is true, in particular, if $F(x)$ is convex for all $x \in \overline{U}$.

For the results of this paragraph see Seigberg-Skordev [15], Skordev [16], and Fournier-Violette [4].

Let $X_0 = \overline{U}, X_1, \cdots, X_{n+1} = E$ and $F_i : X_i \to X_{i+1}$, $i = 0, 1, \cdots, n$ be a sequence of subsets of E and upper semicontinuous, acyclicvalued maps F_i. The sequence $f = (F_n, \cdots, F_0)$ is said to be an acyclic decomposition of F if $F = F_n \circ F_{n-1} \cdots \circ F_0$. Moreover F is admissible if $FixF = \{x \in \overline{U} : x \in F(x)\}$ is compact in U.

For an admissible map $F : \overline{U} \to E$, which is α-contractive or condensing, and admits an acyclic decomposition $f = (F_n, \cdots, F_0)$ we can define the fixed point index. Although this index depends on the decomposition f, it has nevertheless the following properties.

(4.1) $\mathrm{ind}(U, f, E) = \mathrm{ind}(U, (F_n, \cdots F_{i+1}, F_i \circ F_{i-1}, F_{i-2}, \cdots, F_0), E)$ provided that $F_i \circ F_{i-1}$ is still acyclic;

(4.2) If $U = U_1 \cup U_2$ where U_1, U_2 are open in U such that $Fix(F) \cap U_1 \cap U_2 = \emptyset$, then

$$\mathrm{ind}(U, f, E) = \mathrm{ind}(U_1, f, E) + \mathrm{ind}(U_2, f, E)$$

(4.3) If $\mathrm{ind}(U, f, E) \neq 0$ then $FixF \neq \emptyset$;

(4.4) If $H : \overline{U} \times I \to E$ has an acyclic decomposition $h = (H_n, \cdots, H_0)$, is uppersemicontinuous condensing and with compact values, and if H is admissible (i.e., H_t has no fixed points on ∂U), then $\mathrm{ind}(U, h_0, E) = \mathrm{ind}(U, h_1, E)$ where $h_t = (H_n, \cdots, H_{0_t}$ and $H_{0_t}(x) = H_0(x, t)$.

(4.5) (mod$p$) If $y \in F^i(x)$, $x \in F^{p-i}(y)$ and $x \in U$ implies that $y \in U$; and if F and F^p have no fixed points on ∂U, then $\mathrm{ind}(U, f, E) \equiv \mathrm{ind}(U, f^p, E)$ (mod$p$).

In the rare cases when the index does not depend on the decomposition we shall denote it by $\mathrm{ind}(U, F, E)$.

5. The multivalued case

The proof of our first result for multivalued maps is essentially the same as for the singlevalued case (see Theorem 3.1 and Remark 3.1).

Theorem. *Let $F : \overline{\Omega} \to E$ be a compact, upper continuous composition of acyclic maps such that*

(5.1) *$F(\partial\Omega) \subset E \backslash co\Omega$ and F has no fixed points on $\partial\Omega \cap \partial co\Omega$.*

Then

(5.2) $\text{ind}(E, F, \Omega) = 0$.

For α-contractions and condensing maps the techniques of section 3 cannot be carried through to the multivalued case, since they depend on the closedness (openess) of the inverse image of closed (open) sets under the multivalued map $F : \overline{\Omega} \to E$. We could assume the following additional properties:

1. if $x \in F(x)$ (i.e., x is a fixed point of F), then $F^n(x) \subset \Omega$ for all $n \geq 0$;
2. if $F^n(x) \cap \partial\Omega \neq \emptyset$, then $F^{n+1}(x) \cap \overline{\Omega} = \emptyset$, where of course, $F^{n+1}(x) = F(F^n(x) \cap \overline{\Omega})$.

Under these assumptions our methods can be extended to the multivalued case, because we gain enough control on the fixed points of F^n belonging to ∂V_n (here $V_n = F^{1-n}(\Omega)$, with $F^{-1}(\Omega) = \{x : F(x) \subset \Omega\}$). However the conditions listed above are restrictive, although they are verified in the singlevalued case.

Let us do first an easy case, with Ω still a bounded open neighborhood of the origin.

Theorem 5.2. *Let $F : \overline{\Omega} \to E$ be a condensing upper semicontinuous acyclic map. Assume that Ω is convex and*

$$\overline{F(\partial\Omega)} \subset E \backslash \overline{co}\Omega. \tag{5.3}$$

Then

$$\text{ind}(E, F, \Omega) = 0. \tag{5.4}$$

Proof. Let $F_n \cdots F_0$ be an acyclic decomposition of F and let us prove that $\text{ind}(E, (F_n, \cdots, F_0), \Omega) = 0$.

Since $co\Omega$ is a convex neighborhood of 0, the radial retraction π on $\overline{co}\Omega$ exists and is continuous.

The extension of F, $F \circ \pi : E \to E$, coincides with F on $\overline{\Omega}$. Moreover $F \circ \pi(E \backslash \Omega) \subset \overline{F(\partial\Omega)} \subset N_h(0) \backslash \Omega$ for some h such that $\overline{\Omega} \subset N_h(0)$. Then by normality and because $N_h(0) \backslash \Omega$ is acyclic, we have $\text{ind}(N_h(0) \backslash \Omega, (F_n, \cdots, F_0, \pi), E) =$

$\Lambda(F_n, \cdots, F_0, \pi) = 1$. Similarly $\operatorname{ind}(N_h(0), (F_n, \cdots, F_0, \pi), E) = 1$. By additivity we get

$$\operatorname{ind}(\Omega, (F_n, \cdots, F_0), E) = \operatorname{ind}(\Omega, (F_n, \cdots, F_0, \pi), E) = 0.$$

Corollary 5.1. *Let $F : \overline{\Omega} \to E$ be an upper semicontinuous composition of acyclic maps. Assume that F is α-contractive with constant $p < 1$ and it satisfies (5.1). Then*

$$\operatorname{ind}(E, F, \Omega) = 0$$

provided that Ω is convex.

Proof. Let $q > 1$ be such that $qp < 1$. Then the map $G = qF$ satisfies (5.3) and so its index in 0 on Ω. Since G is homotopic to F, via an admissible homotopy, we get the conclusion.

Corollary 5.2. *Let Ω be a convex, open neighborhood of 0 and $F : \overline{\Omega} \to E$ be an upper semicontinuous composition of acyclic maps. Assume that F is an α-contraction with constant $p < 1$, and that*

$$\delta = \inf\{||y|| y \in F(x) \text{ and } x \in \partial\Omega\} > p\rho = p \sup\{||x|| : x \in \partial\Omega\}; \tag{5.5}$$

$$x \notin tF(x) \quad \forall t \geq 1 \quad \forall x \in \partial\Omega. \tag{5.6}$$

Then $\operatorname{ind}(E, F, \Omega) = 0$.

Proof. By (5.3) we can find $\varepsilon > 0$ such that $p + \varepsilon < 1$ and $\rho(p + \varepsilon) < \delta$. Then, the map qF, $q = \frac{p+\varepsilon}{p}$ is homotopic to F by an admissible homotopy and satisfies the conditions of Corollary 5.1.

All results listed above can be readily extended to wedges.

We now examine the case when Ω is not convex, but $F : \overline{\Omega} \to E$ can be extended to $\overline{co\Omega} \backslash \overline{\Omega}$ without introducing new fixed points. This extension is possible when, for example, Ω is star-shaped with respect to the origin, but here we prefer to focus our attention to the case when Ω is a subset of a special wedge.

Recall that a wedge W is special if there exists $y \in W$ such that

$$||x + \lambda y|| \geq ||x|| \tag{5.7}$$

for every $x \in W$ and all $\lambda \geq 0$. Any cone is a special wedge [9]. Notice that for every $k \geq 0$, the vector ky satisfies (5.7) as well. Therefore, without loss of generality we may assume that $||y|| = 1$.

Theorem 5.3. *Let Ω be a bounded open neighborhood of 0 in a special wedge W, and let $F : \overline{\Omega} \to W$ be an upper semicontinuous composition of acyclicvalued maps. Assume that F is an α-contraction with constant $p < 1$ and that (5.5), (5.6) are satisfied. Then*

$$\operatorname{ind}(W, F, \Omega) = 0. \tag{5.8}$$

Proof. As in Corollary 5.2 we have

$$\operatorname{ind}(W, F, \Omega) = \operatorname{ind}(W, G, \Omega)$$

where $G = qF$, $q = (p+\varepsilon)p^{-1}$. Thus there exists $r > 0$ such that $G(\partial\Omega) \subset W \backslash D(0,r)$ and $\overline{\Omega} \subset D(0,r)$. Moreover, since G is upper semicontinuous and W is normal, there exist two open sets V, U such that $0 \in V \subset \overline{V} \subset U \subset \overline{U} \subset \Omega$ and $G^{-1}(W \backslash D(0,r)) \supset \overline{\Omega} \backslash V$.

Let $v, u : W \to [0,1]$ be continuous and such that

$$\begin{aligned} v(x) &= 0 \text{ if } x \in \overline{V} \\ v(x) &= 1 \text{ if } x \in W \backslash U \\ u(x) &= 1 \text{ if } x \in \overline{U} \\ u(x) &= 0 \text{ if } x \in W \backslash \Omega \end{aligned}$$

Since F is an α-contraction and $\overline{\Omega}$ is bounded, there exists $M > 0$ such that $\operatorname{Im} G$ is bounded by M. Choose $b > M + r + 1$ and define

$$H(x) = \begin{cases} G(x) + v(x)by & \text{if } x \in \overline{U} \\ u(x)G(x) + by & \text{if } x \in \overline{\Omega} \backslash U \\ by & \text{if } x \in W \backslash \Omega \end{cases}$$

The map H is upper semicontinuous, is α-contractive with constant p and its restriction to V coincides with G. Moreover $H(D(0,r) \backslash V) \subset W \backslash D(0,r)$. In fact for $z \in G(x)$ and $x \in \overline{\Omega} \backslash V$ we have

$$||z + v(x)by|| \geq ||z|| > r \text{ if } x \in \overline{U}$$

and

$$||u(x)z + by|| \geq b - ||z|| \geq b - M > r + 1 > r \text{ if } x \in \overline{\Omega} \backslash U.$$

Thus

$$\mathrm{ind}(W, G, \Omega) = \mathrm{ind}(W, G, V) = \mathrm{ind}(W, H, V) = \mathrm{ind}(W, H, D(0, r)) = 0.$$

Theorem 5.3 is stated and proved here for α-contractions. Its generalization to condensing maps is still an open question. We therefore see that the results we obtained in the multivalued case are not as general as the ones proved for singlevalued maps.

Acknowledgements. We thank I. Massabo for several helpful conversations in the subject matter of this paper.

References

[1] F. Browder-Yu Qing Yu, Boundary conditions for condensing maps, *Nonlinear Analysis*, TMA 8 (1984), 209-219.

[2] G. Darbo, Punti uniti in trasformazioni a codominio non compatto, *Rend. Sem. Mat. Univ. Padova*, 24 (1955), 353-367.

[3] G. Fournier-M. Martelli, Set valued transformations on the unit sphere, Letters in Math. Physics, 10 (1984), 125-134.

[4] G. Fournier-D. Violette, A fixed point index for compositions of acyclic multivalued maps in Banach spaces, *Operator Equations and Fixed Point Theorems*, S.P. Singh, V.M. Sehgal, J.H.W. Burry, eds., the MSRI-Korea Publications, No. 1, (1986), 139-158.

[5] M. Furi-M. Martelli, A Lefschetz type theorem for the minimal displacement of points under maps defined on a class of ANR's, *Boll. Un. Matem. Ital.* (4) 10 (1974), 174-181.

[6] M. Furi-A. Vignoli, On a property of the unit sphere in a linear normed space, *Bull. Ac. Pol. Sci.*, 18 (1970), 333-334.

[7] M.A. Krasnosel'skii-P.P. Zabreiko, Iterations of operators and fixed points, *Dok. Akad. Nauk SSSR*, 196 (1971), 1006-1009.

[8] C. Kuratowski, Sur les espaces completes, *Fund. Math.* 15 (1930), 301-309.

[9] M. Martelli, Positive eigenvectors of wedge maps, *Annali di Matematica Pura e Applicata*, 4, 145 (1986), 1-32.

[10] I. Massabo-Ch. Stuart, Positive eigenvectors of k-set contractions, *Nonlinear Analysis*, T.M.A. 3 (1979), 35-44.

11] P. Massat, Some properties of condensing mappings, *Ann. Mat. Pura Appl.* (4), 125 (1980), 101-115.

[12] R. Nussbaum, Eigenvectors of nonlinear positive operators and the linear Krein-Rutman theorem, in *Fixed Point Theory, Lectures Notes in Mathematics*, 886, E. Fadell and G. Fournier eds., (1981), 309-331.

[13] W. Petryshyn, On the solvability of $x \in T(x) + \lambda F(x)$ in quasinormal cones with T and F k-set-contractive, *Nonlinear Analysis*, T.M.A. 5 (1981), 585-591.

[14] H.H. Schaefer, *Topological Vector Spaces*, Springer-Verlag GTM, 3, 1970.

[15] H.W. Siegberg-G. Skordev, Fixed point index and chain approximations, *Pacific J. Math.* 102 (1982), 455-486.

[16] G. Skordev, Fixed point index for open sets in Euclidean spaces, Report 45 (1981), *Forschungschwerpunkt Dynamische Systeme*, Universität Bremen.

[17] H. Steinlein, Uber die verallgemeinerten Fixpunktindizes von Itiererten verdichtender Abbildungen, *Man. Math.*, 8 (1973), 251-266.

Recent Progress in Symplectic Geometry

H. Hofer* Rutgers University

I. INTRODUCTION

Since the end of the seventies dramatic progress has been achieved in Hamiltonian dynamics and symplectic geometry. In this survey we shall look at several topics which play a key role in this development. We are however far from being complete and we refer the reader to [B], [Gr3] and [Ra–Am–Ek–Ze] for other aspects. The topics are presented here in their logical, rather than their chronological order. Before we go into details however we give a short "historical" sketch.

The breakthrough in the variational theory of finding periodic solutions of Hamiltonian systems (if we forget the more special geodesic problem) is due to Rabinowitz and Weinstein in '78, [Ra1], [We4]. In his paper Rabinowitz showed that the well known classical variational principle can be effectively used to show the existence of periodic solutions for Hamiltonian systems. This came as a big surprise since this principle was considered as very degenerate and as useless for existence questions (see for example J. Moser's remark in [Mo2]). Motivated by Rabinowitz's result there was a flood of papers studying different types of Hamiltonians, see [Ra3], [Ze1] and generally [Ra–Am–Ek–Ze] for a survey. Moreover it led to the study of new classes of abstract functionals, see [Be–Ra2], [H1]. On the other hand the dual variational principle was introduced in [Cl1] and led to the celebrated Ekland–Lasry result [E–L], Ekelands Morse–theory for convex Hamiltonian systems [E1], the solution of the minimal period problem in [E–H1] for convex Hamiltonian after preliminary results by Ambrosetti and Mancini [A–M] and Clarke–Ekeland [Cl–E] and recently in [E–H3] to global and local symplectic invariants of convex Hamiltonian energy surfaces and their periodic trajectories. A very recent breakthrough is due to C. Viterbo [Vi] who proves the so-called Weinstein conjecture in the $\mathbb{R}^{2n}$ case. This conjecture is concerned with the

* Research partially supported by NSF Grant DMS–8603149
Alfred P. Sloan Foundation and a Rutgers Trustee's Research
Fellowship grant.

existence of periodic solutions for a Hamiltonian system on a prescribed energy surface. Motivated by [Vi] in [H–Ze] a new phenomenon was detected – The Almost–Existence–Mechanism. In [H–Vi] and [F–H–V] more cases of the Weinstein conjecture have been solved. The method of proof in [F–H–V] is particularly interesting since one uses a new ingredient: first order elliptic systems. This brings us to symplectic geometry and Gromov's key contribution to this subject.

Symplectic geometry had developed separately from Hamiltonian dynamics. Though a global symplectic geometry had been conjectured in '65 by Arnold [Ar1], [Ar2], the results stayed local for a long time, see [We1], [Mo1]. In fact embedding results due to Gromov, see [Gr2,3] indicated a high flexibility of the symplectic notion. In '82 Conley and Zehnder [Co–Ze1] succeeded in proving one of the Arnold conjectures using methods from the variational theory of Hamiltonian dynamics and the homotopy index of C. Conley [Co]. Using this "hook–up" M. Chaperon proved a Lagrangian intersection result for the standard torus, [Ch], and in [H2] the most general case for Lagrangian intersection theory in cotangent bundles of compact manifolds was established, see [L–S] for a simpler proof. [Co–Ze1] also inspired A. Floer and J. Sikorav in proving some of the Arnold conjectures in certain spaces [Fl1], [Si]. A very important contribution is then made in '85 by M. Gromov [Gr1]. He introduces his theory of almost holomorphic curves. A. Floer quickly realized that Gromov's (nonvariational) ideas can be connected with the "classical" variational approach used in [Co–Ze1], [H2], [Fl1], [Si]. This leads to Floer's Morse theory for Lagrangian intersections [F2–F5] and then to the Lusternik-Schnirelman theory, simultaneously developed in [F7] and [H3]. A phenomenon well known in the theory of harmonic maps and detected in this set-up by Sacks and Uhbenbeck [S–U] and for holomorphic maps in [Gr1], namely the bubbling off of harmonic or holomorphic spheres plays a key technical role and causes difficulties if certain topological conditions are not met. Recently Floer obtained some very interesting results in this direction [Fl6].

In another direction, in [E–H5], the local rigidity of symplectic maps has been studied, giving a new phenomenon. Rigidity in symplectic geometry had only been accepted and expected as a global phenomenon, see remarks in [B] and

[Gr3] concerning the celebrated Eliashberg Gromov result which says in a variant that the symplectic diffeomorphism group of a compact symplectic manifold is C^0–closed in the diffeomorphism group. Very interestingly the results in [E–H5] show a close relationship between symplectic embedding problems and Hamiltonian dynamics on energy surfaces.

In this survey we shall start with the local rigidity property of symplectic geometry, connect it with Hamiltonian dynamics on an energy surface and study the Weinstein conjecture. Finally, we look at the Arnold conjecture and survey the recent results and explain some of the key technical ingredients.

II. LOCAL RIGIDITY IN SYMPLECTIC GEOMETRY

We denote by (V,ω) the standard symplectic vector space $\mathbb{R}^{2n}$ equipped with the symplectic form $\omega = \langle J\cdot,\cdot\rangle$, where

$$J = \begin{bmatrix} 0 & -I \\ I & 0 \end{bmatrix} \in \mathscr{L}(V)$$

and $\langle\cdot,\cdot\rangle$ denotes the standard inner product. We call a linear map $\phi\colon V \to V$ symplectic if it preserves ω, i.e.

$$\phi^*\omega = \omega, \ (\phi^*\omega)(\xi,\eta) = \omega(\phi(\xi), \ \phi(\eta))$$

We denote by $Sp(V)$ the group of symplectic linear isomorphisms. Given $\phi \in Sp(V)$ we have $\phi^*\omega^n = \omega^n$ so that $\det(\phi) = 1$. Given a smooth map $h\colon U \to V$, defined on an open subset U of V we write $h^*\omega$ for the 2–form on U defined by

$$(h^*\omega)(x)(\xi,\eta) = \omega(h'(x)\xi,h'(x)\eta)$$

where $h'(x)$ is the derivative at $x \in U$. A map h as described above is called symplectic iff $h^*\omega = \omega$ on U. Assume now U_1 and U_2 are open subsets of V. We study the question, when is there a symplectic embedding $h\colon U_1 \to U_2$? By the preceding discussion h will be volume preserving so that the existence of such an embedding implies $\text{vol}(U_2) \geq \text{vol}(U_1)$. A volume condition in V is a "2n–dimensional" condition in contrast to the 2–dimensional condition of being symplectic. Can we say more than $\text{vol}(U_2) \geq \text{vol}(U_1)$ (if $n \geq 2$)? For $r \in [0,+\infty]$ denote by $B(r)$ the open Euclidean ball in V of radius r, where we have $B(0) = \emptyset$ and $B(\infty) = V$. By $\Sigma(r)$ we denote the open symplectic cylinder of radius r, i.e.

$$\Sigma(r) = \{x \in V \mid x_1^2 + x_{n+1}^2 < r^2\}.$$

We ask now the simple-looking question when can $B(r)$ be symplectically embedded into $\Sigma(r')$. Obviously an embedding (even linear) is possible if $r' = r$. Can we perhaps squeeze $B(r)$ into some $\Sigma(r')$ with $r' < r$? It turns out that this is a very deep question. In [Gr1] is proved the following result (among many other things).

THEOREM 1 If there exists a symplectic embedding $h\colon B(r) \to \Sigma(r')$ then $r' \geq r$.

This result is proved using nonlinear first order elliptic systems (Cauchy–Riemann type operators). Motivated by Gromov's result the notion of a symplectic capacity function has been introduced in [E–H5]. The idea is to put the embedding problem $h\colon U_1 \to U_2$ on an axiomatic foundation. This approach actually leads to an interesting hook-up with the fixed energy problem in Hamiltonian dynamics and in particular this connection can be used to give an alternative proof of Theorem 1 using periodic solution of Hamiltonian systems on a prescribed energy surface, see [E–H5] for more details.

1. SYMPLECTIC CAPACITY FUNCTIONS

We denote by $\wp$ the power set of V.

Definition 1 A map $\alpha: \wp \to [0,+\infty]$ satisfying the following three axioms is called a symplectic capacity function

(A1) (normalisation) $\alpha(\Sigma(1)) = \alpha(B(1)) = \pi$
(A2) (monotoncity) $S \subset T \Rightarrow \alpha(S) \leq \alpha(T)$
(A3) (conformality) If $\psi: V \to V$ is a diffeomorphism satisfying $\psi^* \omega = c\,\omega$ for some $c > 0$ then, $\alpha(\psi(S)) = c\,\alpha(S)$ for $S \in \wp$.

It is not difficult to show that the existence of a symplectic capacity function α is equivalent to Gromov's result. Let us define, with D being the symplectic diffeomorphism group of V,

$$(1)\quad \hat{\alpha}(S) = \inf\{\pi r^2 \in [0,+\infty] \mid \text{there exists } \psi \in D \text{ with } \psi(S) \subset \Sigma(r)\}$$
$$\check{\alpha}(S) = \sup\{\pi r^2 \in [0,+\infty] \mid \text{there exists } \Psi \in D \text{ with } \psi(B(r)) \subset S\}.$$

An easy theorem is

THEOREM 2 $\hat{\alpha}$ and $\check{\alpha}$ are symplectic capacity functions satisfying $\check{\alpha} \leq \hat{\alpha}$. Moreover given any symplectic capacity function α we have

$$\check{\alpha} \leq \alpha \leq \hat{\alpha}.$$

So $\check{\alpha}$ is the minimal and $\hat{\alpha}$ the maximal symplectic capacity function.

We call a subset S of V a linear ellipsoid if there exists a quadratic form q, with $q(x) \geq 0$, $x \in V$, such that

$$S = \{x \in V \mid q(x) < 1\} =: S_q$$

The collection of all linear ellipsoids will be denoted by $\wp_{LE}$. An ellipsoid is a set S such that

$$S = \psi(T)$$

for some $\psi \in D$ and $T \in \varphi_{LE}$. φ_E denotes the collection of all ellipsoids. Note that in "short hand"

$$\varphi_E = D \circ \varphi_{LE}$$

For $o < r_1 \leq r_2 \ldots \leq r_1$ define with $r = (r_1, \ldots r_n)$ $S(r) \in \varphi$ by

$$S(r) = \{x \in V \mid \sum_{i=1}^{n} \frac{1}{r_i^2} (x_i^2 + x_{i+n}^2) < 1\}.$$

It is a standard result in linear symplectic geometry, see [Ar2], that given any $S \in \varphi_{LE}$ which is bounded there exists r and $\psi \in Sp(V)$ such that

$$\psi(S) = S(r)$$

Using this fact one easily verifies that

$$\hat{\alpha}|\varphi_E = \check{\alpha}|\varphi_E$$

So there is only one symplectic capacity function on φ_E. This leads to a good open problem.

PROBLEM 1 What is the maximal class φ_M of φ such that $\hat{\alpha}|\varphi_M = \check{\alpha}|\varphi_M$? A result which we will obtain later suggest that if S is symplectomorphic to an open convex set, then $\hat{\alpha}(S) = \check{\alpha}(S)$.

Given a symplectic capacity function α we can study homeomorphisms in V preserving α. We start in the linear category.

DEFINITION 2 By $A(V)$ we denote the subgroup of $GL(V)$ consisting of all maps ϕ such that

$$(2) \quad \alpha(\phi(S)) = \alpha(S)$$

for every bounded $S \in \wp_{LE}$.

Note that the definition does not depend on the choice of α since on $\wp_{LE}$ $\alpha = \hat{\alpha} = \alpha$.

The key linear result is

THEOREM 3 $A(V) = Sp(V) \cup \Gamma \circ Sp(V)$, where Γ is a linear isomorphism $(V,\omega) \to (V,-\omega)$.

For the proof which is basically straightforward but somewhat tedious we refer the reader to [E–H5].

A useful result which we need later on is

PROPOSITION 1 If ϕ: $V \to V$ is linear and $\alpha(\phi(S)) > 0$ for every nonempty bounded $S \in \wp_{LE}$, then $\phi \in GL(V)$.

Next we move to the nonlinear category.

2. A LOCAL RIGIDITY THEOREM AND THE ELIASHBERG–GROMOV–THEOREM

Having a symplectic capacity function and the key linear result Theorem 3 we start with a technical result.

THEOREM 4 Assume (h_k): $B(1) \to V$ is a sequence of continuous maps converging uniformly to a continuous map h: $B(1) \to V$ which is differentiable at 0 with derivative $h'(0)$. Suppose

$$\alpha(h_k(S)) = \alpha(S) \ , \ k \in \mathbb{N}$$

for all $S \in \varphi_{LE}$, $S \subset B(1)$. Then $h'(0) \in A = A(V)$.

An easy corollary of Theorem 4 is the result we are aiming at.

THEOREM 5 (Local symplectic rigidity) Assume $(h_k): B(1) \to V$ is a sequence of symplectic embeddings converging uniformly to a continuous map $h: B(1) \to V$ which is differentiable at 0 with derivative $h'(0)$. Then $h'(0) \in Sp(V)$.

Assuming Theorem 4 for the moment we know since the h_k are α-preserving that $h'(0) \in A$. If n is odd then Γ is orientation reversing. Since $h'(0)$ has to be orientation preserving we must have $h'(0) \in Sp(V)$. If n is even consider the symplectic vector space $(V \oplus \mathbb{R}^2, \omega_V \oplus \omega_{\mathbb{R}^2})$ which is symplectic isomorphic to $(\mathbb{R}^{2n+2}, \omega_{\mathbb{R}^{2n+2}})$, say by a map γ. Define $\tilde{h}_k$ by

$$\tilde{h}_k = \gamma \circ (h_k \times I_{\mathbb{R}^2}) \circ \gamma^{-1}$$

Then $\tilde{h}_k$ is a symplectic embedding and $\tilde{h}_k$ converges uniformly to $\tilde{h}$ defined by

$$\tilde{h} = \gamma \circ (h \times I_{h^2 2}) \circ \gamma^{-1}$$

Now $n + 1$ is odd and $\tilde{h}'(0) \in Sp(\mathbb{R}^{2n+2})$ by the previous argument. Since

$$\tilde{h}'(0) = \gamma \circ (h'(0) \times I_{\mathbb{R}^2}) \circ \gamma^{-1}$$

we see that $h'(0) \times I_{\mathbb{R}^2}$ is symplectic which is only possible if

$h'(0) \in Sp(V)$.

Next we sketch a proof of Theorem 4.

Given $S \in \varphi_{LE}$, $S \subset B(1)$, and $\epsilon > 0$ we find $T \in \varphi_E$ such that

$$cl(h(S)) \subset T \ , \ \hat{\alpha}(T) \leq \hat{\alpha}(h(S)) + \epsilon$$

Since for k large enough we have $h_k(S) \subset T$ we see that

$$\begin{aligned}\hat{\alpha}(S) &= \alpha(S) \\ &= \alpha(h_k(S)) \\ &\leq \hat{\alpha}(h_k(S)) \\ &\leq \hat{\alpha}\ (T) \\ &\leq \hat{\alpha}\ (h(S)) + \epsilon.\end{aligned}$$

$\epsilon > 0$ being arbitrary implies

$$(1) \quad \hat{\alpha}(S) \leq \hat{\alpha}(h(S))$$

For $t \in (0,1)$ using (1) we compute

$$\begin{aligned}(2) \quad \hat{\alpha}(S) &= \hat{\alpha}(\tfrac{1}{t}(tS)) \\ &= \frac{1}{t^2}\ \hat{\alpha}(tS) \\ &\leq \frac{1}{t^2}\ \hat{\alpha}(h(tS)) \\ &= \hat{\alpha}(\tfrac{1}{t}h(tS))\end{aligned}$$

If now $t \downarrow 0$ the maps $\frac{1}{t}h(t\cdot)$ converge uniformly to $h'(0)$. Using the same arguments as above we see that

$$(3) \quad \limsup_{t \downarrow 0}\ \hat{\alpha}(\tfrac{1}{t}h(tS)) \leq \hat{\alpha}(h'(0)S)$$

combining (2) and (3) gives

$$(4) \quad \hat{\alpha}(S) \leq \hat{\alpha}(h'(0)S)$$

for every $S \in \varphi_{LE}$ with $S \subset B(1)$. By Proposition 1 this implies that $h'(0) \in GL(V)$. Next we have to show the reverse inequality in (4). Since h is differentiable at 0 we find a nondecreasing map $\epsilon\colon (0,1) \to (0,+\infty)$ such that $\epsilon(s) \to 0$ as $s \to 0$ and assuming $h(0) = 0$

$$|h(x) - h'(0)x| \leq \epsilon\,(|x|)\;|x|.$$

Given $\delta \in (0,1)$ we find $k(\delta)$ such that for $k \geq k(\delta)$

$$|h_k(x) - h'(0)x| \leq \epsilon(|x|)\;|x| + \delta.$$

Pick $\gamma > 0$. For $\tau \in (0,1)$ small enough we have $(1+\gamma)\tau < 1$ which we assume in the following. Take an arbitray $S \in \varphi_{LE}$ with $S \subset B(1)$. We shall show that

$$(5) \quad h_k((1+\gamma)\tau S) \supset h'(0)(\tau S)$$

for k large enough and τ small enough. (5) will be a simple consequence of a Bouwer degree argument which we leave to the reader. Hence from (5) we infer

$$\begin{aligned}\hat{\alpha}(h'(0)(\tau S)) &= \alpha(h'(0)(\tau S)\\ &\leq \alpha(h_k((1+\gamma)\tau S))\\ &= \alpha((1+\gamma)\tau S)\\ &\leq \hat{\alpha}\;((1+\gamma)\tau S)\\ &\leq (1+\gamma)^2\tau^2\;\hat{\alpha}(S)\end{aligned}$$

Hence we obtain for $S \in \varphi_{LE}$, $S \subset B(1)$

$$(6) \quad \hat{\alpha}(h'(0)S) \leq \hat{\alpha}(S)$$

Now combining (4) and (6) and using that $h'(0)$ is linear we must have

$$(7) \quad \hat{\alpha}(S) = \hat{\alpha}(h'(0)S) \qquad S \in \varphi_{LE} \text{ bounded.}$$

This shows that $h'(0) \in A$.

Finally we give as a corollary the celebrated Eliashberg–Gromov result. The result had been announced in 1981 by Eliashberg, but had never been published. According to Gromov the result had been proved by Eliashberg using a very complicated combinatorial argument based on Poincare's proof of the last geometric theorem. In [Gr2] Gromov sketches a (difficult) proof based on the Nash–Moser implicit function theorem and Theorem 1. Because of its proof the result had been seen as a global phenomenon see [B] and [Gr3]. However the proof used here shows that it is indeed a very local result.

THEOREM 6 (Eliashberg–Gromov) Let (M,ω) be a symplectic manifold. Then the symplectic diffeomorphism group $D_\omega(M)$ is for the C^0–compact open topology closed in the diffeomorphism group Diff(M).

PROOF: Using Darboux charts we can reduce the result to Theorem 5.

3. RELATIONSHIP TO THE FIXED ENERGY PROBLEM

In this part we shall show that there is a relation to the fixed energy problem in Hamiltonian dynamics. This relation is not well understood today and a better understanding could be mutally beneficial for both fields.

Symplectic capacity functions give us an obstruction against embedding one set symplectically into another set. The obstruction is a number $\alpha(S)$ and naively one would guess the obstruction to be visible as a class of 2–dimensional subsets in S. Since, while embedding S into T the problems should arise at the boundary of S, the geometric obstruction set should have a trace on the boundary. The trace of a nice 2–dimensional set should be 1–dimensional and in nice cases it should be a loop on ∂S.

Assuming ∂S to be smooth, there is a class of nice loops on ∂S, namely periodic Hamiltonian trajectories. To be more precise we need some notation.

Given a connected compact smooth hypersurface Λ in V the normal bundle is trivial as a consequence of Alexander duality, [Sp]. Moreover $V\backslash\Lambda$ has a bounded and an unbounded component. We shall denote the bounded component by B_Λ. There exists a canonical one–dimensional distribution $\mathscr{L}_\Lambda \to \Lambda$, $\mathscr{L}_\Lambda \subset T\Lambda$, defined by

$$\mathscr{L}_\Lambda = \{(x,\xi) \in T\Lambda \mid \xi \perp_\omega T_x\Lambda\}$$

$\mathscr{L}_\Lambda \to \Lambda$ is called the characteristic distribution.

DEFINITION 3 A closed characteristic or periodic Hamilitonian trajectory on Λ is a submanifold P of Λ diffeomorphic to S^1 such that $TP = \mathscr{L}_\Lambda | P \cdot \mathscr{P}(\Lambda)$ denotes the collection of all closed characteristic on Λ.

We need further the definition of contact type, [We3], and restricted contact type, [E–H4].

DEFINITION 4 A compact connected smooth hypersurface $\Lambda \subset V$ is said to be of contact type if there exists a 1–form λ on Λ such that $\lambda(x,\xi) \neq 0$ for $(x,\xi) \in \mathscr{L}_\Lambda$ and $\xi \neq 0$ so that $d\lambda = \omega/\Lambda$. If λ can be extended to V such that $d\lambda = \omega$ we say Λ is of restricted contact type. The collection of all contact type hypersurfaces in V will be denoted by ℓ, the subcollection of restricted contact type by ℓ_r. One has the following result

THEOREM 7 There exists a symplectic capacity function α such that for $\Lambda \in \ell_r$ there exists a positive integer k and a closed characteristic

$P \in \mathcal{P}(\Lambda)$ with

$$(1) \quad \alpha(B_\Lambda) = k|\int \lambda|P \;|$$

Moreover for every $S \in \varphi$ we have

$$\alpha(S) = \inf \{\alpha(U) | \; U \text{ open } \; U \supset S\}$$

and for every nonempty open U

$$\alpha(U) = \sup \{\alpha(B_\Lambda) \mid B_\Lambda \subset U \,,\, \Lambda \in \ell_r\}.$$

Some ideas concerning the proof of Theorem 7 will be given in III. 4.

Here are some open problems.

PROBLEM 2 Can we pick $P \in \mathcal{P}(\Lambda)$ such that

$$\alpha(B_\Lambda) = | \int \lambda|P \;|$$

i.e can we take $k = 1$? If B_Λ is symplectomorphic to a convex domain this is actually possible.

PROBLEM 3 Assume $\Lambda \in \ell_r$ is homeomorphic to S^{2n-1}. Is it true that there exists $\psi \in D$ such that

$$\psi(B_\Lambda) \subset \Sigma(\hat{\alpha}(B_\Lambda)) \;?$$

If yes, is it true that $\psi(\Lambda) \cap \partial\Sigma(\hat{\alpha}(B_\Lambda))$ contains a closed characterstic on $\psi(\Lambda)$?

III. THE FIXED ENERGY PROBLEM IN HAMILTONIAN DYNAMICS AND THE WEINSTEIN CONJECTURE.

As already mentioned in the introduction the breakthrough in questions of existence of periodic solutions on a prescribed energy surface is due to Rabinowitz. Motivated by some further results by Rabinowitz, Weinstein formulated a conjecture which carries his name. A very recent breakthrough in proving the Weinstein conjecture is due to C. Viterbo, [Vi]. Here we shall start with an almost existence phenomenon which had been detected by E. Zehnder and the author and which was motivated by Viterbo's Theorem, [H–Ze]. Moreover we shall survey strong extensions of these results to the cotangent bundle case [Ho–Vi] and the $P \times \mathbb{C}^\ell$ case [F–H–V], where P is a compact symplectic manifold. Then we indicate how one can refine the method to tackle some new kind of fixed point theorem, [E–H4]. Finally we give an idea how a symplectic capacity function based as the fixed energy problem can be constructed.

1. AN ALMOST EXISTENCE RESULT FOR PERIODIC SOLUTIONS

We use the notation as introduced in II.3.

DEFINITION 5 Given a connected compact smooth hypersurface Λ in V we call a diffeomorphism $\psi\colon (-1,1) \times \Lambda \to U$ onto an open bounded neighborhood U of Λ in V a parametrized family of compact hypersurfaces modeled on Λ provided $\psi(0,x) = x$ for every $x \in \Lambda$. We also put $\Lambda_\epsilon = \psi(\{\epsilon\} \times \Lambda)$ and we shall write (Λ_ϵ) or $(\Lambda_\epsilon)_{\epsilon\in(-1,1)}$ instead of ψ.

Denote by λ a 1–form on V such that $d\lambda = \omega$ and write $a(P)$ for $|\int \lambda|P|$ where $P \in \mathcal{P}(\Lambda)$. We have the following surprising almost existence result for periodic solutions, see [H–Ze].

THEOREM 8 (Almost existence) Given $\mathcal{F} = (\Lambda_\epsilon)$ where $\Lambda = \Lambda_0$ is a compact connected hypersurface in V there exists a constant $d = d(\mathcal{F}) > 0$ such that for given $\delta \in (0,1)$ there exists an $|\epsilon| < \delta$ so that the following is true

$$\mathcal{P}(\Lambda_\epsilon) \neq \emptyset \text{ and there exists } P_\epsilon \in \mathcal{P}(\Lambda_\epsilon) \text{ such that } a(P_\epsilon) \leq d.$$

Theorem 8 has at least two interesting consequences. Assume first Λ is of contact type. As it was shown in [We3] there exists a vectorfield η defined on an open neighborhood of Λ in V which is transverse to Λ so that $L_\eta \omega = \omega$. Using the flow associated to η we can construct a ψ: $(-1,1) \times \Lambda \to U$ such that

$$(1) \quad T\psi_\epsilon(\mathcal{L}_\Lambda) = \mathcal{L}_{\Lambda_\epsilon}$$

where $\psi_\epsilon: \Lambda \to \Lambda_\epsilon : \psi_\epsilon(x) := \psi(\epsilon,x)$. Cleary (1) implies

$$(2) \quad \psi_\epsilon(\mathcal{P}(\Lambda)) = \mathcal{P}(\Lambda_\epsilon)$$

Since by Theorem 8 there is a $|\epsilon_0| < 1$ with $\mathcal{P}(\Lambda_{\epsilon_0}) \neq \emptyset$ we see that $\mathcal{P}(\Lambda) \neq \emptyset$. This proves

THEOREM 9 (Viterbo) If Λ is of contact type then $\mathcal{P}(\Lambda) \neq \emptyset$.

Theorem 9 is a generalization of the Weinstein conjecture.

DEFINITION 6 We say a compact smooth hypersurface Λ admits an a priori estimate if there is a family (Λ_ϵ) as in definition 5, $\Lambda_0 = \Lambda$, and a constant $c > 0$ such that

$$(3) \quad \ell(P) \leq \frac{1}{c}\, a(P)$$

for every $P \in \bigcup_{\epsilon\in(-1,1)} \mathscr{P}(\Lambda_\epsilon)$. It is of course not assumed that $\bigcup \mathscr{P}(\Lambda_\epsilon) \neq \emptyset$. As a consequence of the almost existence result we have

THEOREM 10 ([H–Ze]) If Λ admits an a priori estimate then $\mathscr{P}(\Lambda) \neq \emptyset$.

This follows immediately from Theorem 8 over the Ascoli–Arzela Theorem. Theorem 10 had been (vaguely) conjectured by P. Rabinowitz: a priori estimates imply existence. Defintion 6 makes precise what a priori estimate means.

We sketch now a proof of the almost existence result. The idea is to construct a Hamiltonian system $H: V \to \mathbb{R}$ such that

$$H^{-1}((0,b)) = \bigcup_{|\epsilon|<\delta} \Lambda_\epsilon$$

for a suitable large number b and

$$H^{-1}(c) = \Lambda_{\epsilon(c)} \qquad \text{for} \quad c \in (0,b),$$

so that the following holds: If $x: [0,1]/\{0,1\} \to V$ is a smooth 1–perodic loop satisfying

$$\dot{x} = X_H(x),$$

where X_H is the Hamiltonian vectorfield associated to H via $dH = \omega(X_H,\cdot)$ and if

$$\phi_H(x) = \frac{1}{2}\int_0^1 <-J\dot{x}, x> - \int_0^1 H(x) \quad > 0$$

then $x([0,1]) \subset \Lambda_\epsilon$ for some $|\epsilon| < \delta$. Clearly this implies Theorem 8 up to the a priori bound for the action $a(P)$. So the crucial point is the construction of the Hamiltonian. Fix $\delta \in (0,1)$ and denote by B the bounded component of $V\backslash\psi([-\delta,\delta] \times \Lambda)$ and by A the unbounded component. We may assume $0 \in B$. We define

$$(4) \quad \gamma := \text{diam}\,(U)$$

and fix numbers r,b such that

$$(5) \quad \gamma < r < 2\gamma$$
$$\tfrac{3}{2}\pi r^2 < b < 2\pi r^2$$

Note that γ,r,b do not depend on the choice of δ. Next we pick smooth functions $j\colon (-1,1) \to \mathbb{R}$, $g\colon (0,\infty) \to \mathbb{R}$ satisfying

$$(6) \quad j|(-1,-\delta] = 0,\ j|[\delta,\,1) = b,\ j'(s) > 0 \text{ for } -\delta < s < S$$
$$g(s) = b \quad \text{for} \quad s \le r,\ g(s) = \tfrac{3}{2}\pi s^2 \quad s \text{ large},$$
$$g(s) \ge \tfrac{3}{2}\pi s^2 \quad \text{for} \quad s > r,\ 0 < g'(s) \le 3\pi s \quad \text{for} \quad s > r.$$

Next we define a very special Hamiltonian $H \in C^\infty(V,\mathbb{R})$ by

$$(7) \quad H(x) = \begin{cases} 0 & \text{if } x \in B \\ j(\epsilon) & \text{if } x \in \Lambda_\epsilon,\ -\delta \le \epsilon \le \delta \\ b & \text{if } x \in A^c,\ |x| \le r \\ g(|x|) & \text{if } |x| > r \end{cases}$$

We note that

$$(8) \quad -b + \tfrac{3}{2}\pi\,|x|^2 \le H(x) \le b + \tfrac{3}{2}\pi\,|x|^2$$
$$\text{for} \quad x \in V$$

If $x\colon [0,1]/\{0,1\} \to V$ is a smooth loop we define $\phi(x)$ by

$$(9) \quad \phi(x) = \frac{1}{2}\int_0^1 \langle -J\dot{x}, x\rangle\,dt - \int_0^1 H(x(t))dt$$

We have the following crucial observation

LEMMA 1 Assume x is a 1–periodic solution of the Hamiltonian system $\dot{x} = X_H(x)$ satisfying $\phi(x) > 0$. Then $P: = x([0,1])$ is a closed characteristic in some $\mathscr{P}(\Lambda_\epsilon)$ with $|\epsilon| < \delta$.

The proof is simple. If x is constant we have $\phi(x) \le 0$. If x is nonconstant and $|x(t)| \ge r$ for some t we have $|x(t)| = |x(0)|$. Then x satisfies

$$-J\dot{x} = \frac{g'(|x|)}{|x|} x$$

With $v = |x(0)|$ we compute

$$\begin{aligned}\phi(x) &= \frac{1}{2} g'(v)v - g(v)\\ &\le \frac{3}{2}\pi v^2 - g(v)\\ &\le \frac{3}{2}\pi v^2 - \frac{3}{2}\pi v^2\\ &= 0\end{aligned}$$

Hence we can only have $x([0,1]) \subset \Lambda_\epsilon$ for some $\epsilon \in (-\delta,\delta)$, which immediately implies our assertion.

Now 1–periodic solutions of $\dot{x} = X_H(x)$ can be found as critical points of ϕ on a suitable Hilbert space of loops. In view of Lemma 1 it is enough to find a critical point x of ϕ satisfying $0 < \phi(x) \le b$. In fact if $P = x([0,1])$ we find since $P \in \mathscr{P}(\Lambda_\epsilon)$ for some $|\epsilon| < \delta$

$$\begin{aligned}a(P) &\le \frac{1}{2} \int_0^1 \langle -J\dot{x}, x\rangle dt\\ &= \phi(x) + \int_0^1 H(x)dt\\ &\le b + b\\ &= 2b\end{aligned}$$

Define $d(\mathscr{F}) := 16\pi \operatorname{diam}(U)^2$ $(> 2b)$.

In order to find critical points for ϕ we take a variational set up as in [Be–Ra2]. Denote by $E = H^{\frac{1}{2}}(S',V)$ the Hilbert space of all functions

$x \in L^2(o,1\ ;V)$ satisfying

$$x(t) = \sum_{k \in \mathbb{Z}} \exp(2\pi ktJ)x_k \quad , \quad x_k \in V$$

$$\sum_{k \in \mathbb{Z}} |k|\ |x_k|^2 < \infty$$

As an inner product we take

$$(x,y) = (2\pi \sum |k|\ \langle x_k,y_k\rangle) + \langle x_o,y_o\rangle$$

and define $\|x\| = (x,x)^{\frac{1}{2}}$. We have an orthogonal decomposition $E = E^- \oplus E^o \oplus E^+$, where E^-, E^o, E^+ correspond to the subspaces $k < 0$, $k = 0$, and $k > 0$. We denote by P^-, P^o and P^+ the corresponding orthogonal projections. The functional ϕ previously defined for smooth loops extends to a smooth functional on E and is given by

$$(10) \quad \phi(x) = \frac{1}{2}((-P^- + P^+)x,x) - \int_0^1 H(x(t))dt$$

Using the compact embedding $E \subset L^P$, $1 \leq p < \infty$, we see that the gradient ϕ': $E \to E$ of ϕ is a compact map, which has linear growth since H is quadratic outside a big ball. We have

$$(11) \quad \phi(0) = 0,\ \phi'(0) = 0 \ \text{ and } \ \phi''(0) = -P^- + P^+$$

An easy consequence of (11) is

LEMMA 2 There exist $\alpha \in (0,1)$ and $\beta > 0$ such that

$$\phi|\Gamma \geq \beta \ \text{ where } \ \Gamma = \{x \in E^+ \mid \|x\| = \alpha\}.$$

Define $e(t) = \frac{1}{\sqrt{2\pi}} \exp(2\pi tJ)(1,0,...,0)$ and put $\hat{E} = E^- \oplus E^0 \oplus \mathbb{R}e$. We define a bounded subset Σ of E by

$$\Sigma = \{x = x^- + x^o + se \mid \|x^- + x^o\| \leq \tau \ , s \in [0,\tau]\}.$$

where $\tau > 1$ will be specified later.

LEMMA 3 Denote by $\partial\Sigma$ the boundary of Σ in $\hat{E}$. If $\tau > 1$ is sufficiently large we have $\phi | \partial\Sigma \leq 0$.

This is an immediate consequence of the properties of H.

What we have established so far is the existence of what is called a linking, see [Be–Ra2], [H1]. Now we solve the ordinary differential equation $x' = -\phi'(x)$ in order to obtain a global flow $\mathbb{R} \times E \to E$: $(t,x) \to x \cdot t$. Clearly

$$\frac{d}{dt}\phi(x\cdot t) = - \|\phi'(x\cdot t)\|^2 \leq 0$$

Since $t \to \phi(x\cdot t)$ is a nonincreasing map we see that

$$(\partial\Sigma) \cdot t \cap \Gamma = \emptyset \quad \text{for} \quad t \geq 0$$

Our aim is to show that

$$\Sigma \cdot t \cap \Gamma \neq \emptyset \quad \text{for} \quad t \geq 0$$

So we need

LEMMA 4 $\Sigma \cdot t \cap \Gamma \neq \emptyset$ for $t \geq 0$.

The statement is equivalent to the solvability of

$$(12) \quad \begin{array}{l} (P^- + P^o)(x\cdot t) = 0, \quad \|x\cdot t\| = \alpha \\ t \geq 0 \ , x \in \Sigma \end{array}$$

Applying the variation of constant formula to $x' = -\phi'(x)$ we can write

$$x \cdot t = e^t x^- + x^o + e^{-t} x^+ + B(t,x)$$

where the map $B: \mathbb{R} \times E \to E$ is smooth and maps bounded sets into precompact sets. Hence (12) is equivalent to

$$(13) \quad 0 = x^- + x^0 + (e^{-t}P^- + P^0)\, B(t, x^- + x^0 + se)$$
$$0 = se - (s + \alpha - \|(x^- + x^0 + se) \cdot t\|)e$$

Clearly (13) is of the form

$$(14) \quad 0 = z + T_t(z) \quad z \in \Sigma$$

We know already that $0 \neq z + T_t(z)$ for $z \in \partial\Sigma$ and $t \geq 0$. Since T is compact we can apply Leray–Schauder Degree Theory to find with $\dot{\Sigma} =$ interior of Σ in $\hat{E}$

$$\begin{aligned} &\operatorname{dey}(I + T_1, \dot{\Sigma}, 0) \\ &= \operatorname{dey}(I - \alpha e, \dot{\Sigma}, 0) \\ &= \operatorname{dey}(I, \dot{\Sigma}, \alpha e) \\ &= 1 \end{aligned}$$

since $\alpha e \in \dot{\Sigma}$

LEMMA 5 $c = \lim_{t \to \infty} \sup_{x \in \Sigma} \phi(x \cdot t)$ exists.

Moreover $+\infty > c \geq \beta > 0$ and c is a critical level of ϕ.

PROOF The asymptotic behavior of H implies immediately that ϕ satisfies the Palais–Smale condition, see [Am–Ze], [H–Ze]. Since $\phi(\Sigma)$ is bounded in $\mathbb{R}$ we have $c < +\infty$. Since the map $t \to \sup_{x \in \Sigma} \phi(x \cdot t)$ is nonincreasing c exists.

From Lemma 4 we infer that $\Sigma \cdot t \cap \Gamma \neq \emptyset$ for $t \geq 0$ which implies

$$\sup \phi(\Sigma \cdot t) \geq \inf \phi(\Gamma) \geq \beta > 0$$

Hence $0 < c < +\infty$. Assuming that c is not a critical level we find using the PS–condition that

$$\|\phi'(x)\| \geq \epsilon \quad \text{if} \quad \phi(x) \in [c-\epsilon,\ c+\epsilon]$$

for some $\epsilon > 0$. So there exists $T > 0$ such that $\phi^{c+\epsilon} \cdot T \subset \phi^{c-\epsilon}$, where $\phi^{r} = \phi^{-1}((-\infty,r])$. If $t > 0$ is large enough we have by the definition of c

$$\Sigma \cdot t \subset \phi^{c+\epsilon}$$

Hence

$$\begin{aligned} \Sigma \cdot (t+T) = (\Sigma \cdot t) &\cdot T \\ &\subset \phi^{c+\epsilon} \cdot T \\ &\subset \phi^{c-\epsilon} \end{aligned}$$

showing that $c \leq c - \epsilon$ which is a contradiction.

2. EXTENSIONS TO MORE GENERAL SPACES AND REMARKS

Let us first formulate the general Weinstein conjecture. If we replace (V,ω) by a symplectic manifold (M,ω) and Λ is a compact smooth hypersurface in M we can define again contact type.

CONJECTURE (Weinstein) If Λ is a smooth compact hypersurface of contact type in (M,ω) then $\mathcal{P}(\Lambda) \neq \emptyset$.

Actually Weinstein also assumed that $H^1(\Lambda; \mathbb{R}) = 0$. This hypothesis seems however not to be crucial.

Moreover we can ask when does an almost existence result hold, which is a more general concept then the Weinstein conjecture.

First we consider the contangent bundle case. Let $M = T^*N$ be the cotangent bundle of a compact connected smooth manifold of dimension at least two. Given a connected compact hypersurface $\Lambda \subset M$ we say Λ encloses the zero section if Λ does not intersect the zero section $N \subset T^*N$ and $M\backslash\Lambda$ has exactly two components, so that the bounded component contains N. M is equipped with the standard symplectic form $d\lambda$, where $\lambda =$ "pdq". Again we can define a parametrized family. We have

THEOREM 11 ([H–Vi]) Assume N is a compact connected smooth manifold of dimension at least two. Let $\mathcal{F} = (\Lambda_\epsilon)$ be a parmetrized family modelled on a compact connected hypersurface Λ enclosing the zero section. Then there is a constant $d(\mathcal{F}) > 0$ such that for every $\delta \in (0,1)$ there is a $|\epsilon| < \delta$ with $\mathcal{P}(\Lambda_\epsilon) \neq \emptyset$ containing P_ϵ and

$$0 < | \int \lambda | P_\epsilon | \leq d$$

In particular the Weinstein conjecture holds and the existence mechanism based on a priori estimates.

Another result is the following

THEOREM 12 ([F–H–V]) We have an almost existence result for families (Λ_ϵ) modelled on a compact connected smooth hypersurface $\Lambda \subset P \times \mathbb{C}^\ell (\ell \geq 1)$, with trivial normal bundle. Here $P \times \mathbb{C}^\ell$ is equipped with the symplectic structure $\omega_P \oplus \omega_{\mathbb{C}^\ell}$ and (P, ω_P) is a compact symplectic manifold so that ω_P vanishes on $\pi_2(P)$. $\mathbb{C}^\ell \simeq \mathbb{R}^{2\ell}$ is equipped with the standard structure.

The proof of Theorem 11 is very technical and based on methods developed in [H2] in the proof of one of the Arnold conjectures. Theorem 12 is proved with the aid of first order elliptic systems. This is conceptually much simpler though the technical difficulties are still nontrivial. A proof of Theorem 11 using the machinery of Theorem 12 should be possible.

There is obviously one open problem:

PROBLEM 4 Does there exist a smooth compact hypersurface Λ in V with $\mathcal{P}(\Lambda) = \emptyset$?

3. GENERALIZED FIXED POINT PROBLEMS

An interesting question in the global perturbation theory of Hamiltonian systems is the following. Consider an autonomous Hamiltonian system

$$\text{(US)} \quad -J\dot{x} = H'(x)$$

having $\Lambda = H^{-1}(1)$ as a compact regular energy surface. We call (US) the unperturbed system. Assume now (US) is perturbed between time zero and time one by a nonautonomous Hamiltonian system giving the perturbed equation

$$\text{(PS)} \quad -J\dot{x} = H'(x) + \hat{\epsilon}_t'(x)$$

where $\hat{\epsilon}$ has compact support and $\text{supp}(\hat{\epsilon}) \subset [0,1] \times V$. We are interested in solutions of (PS) which agree before time zero and after time one with the unperturbed movement up to a phase shift. More precisely do there exist solutions x of (US) and y of (PS) and a number $\delta \in \mathbb{R}$ such that

$$\begin{aligned} &H(x(t)) = 1 \quad \text{for all} \quad t \in \mathbb{R} \\ &y(t) = x(t) \quad \text{for all} \quad t \leq 0 \\ &y(t) = x(t+\delta) \quad \text{for} \quad t \geq 1 \end{aligned}$$

If $\mathcal{L}_\Lambda \to \Lambda$, $\Lambda = H^{-1}(1)$, is the characteristic line bundle we denote by $L_\Lambda(x)$ for $x \in \Lambda$ the leaf through x? If ψ is the time–one–map of (PS) (we assume it exists) then the problem described is equivalent to the following generalized fixed point problem

$$\psi(x) \in L_\Lambda(x) \, , \quad x \in \Lambda$$

Here we give two results which can be proved by a tricky extension of the method first described. The details can be found in [E–H4]. See also [Mo1] for local results in this direction.

We denote by G the group $\mathbb{Z}_2$ acting in the usual way by $x \to -x$ on V.

THEOREM 12 Assume Λ is a compact connected smooth hypersurface which is G–invariant and of restricted contact type. Let ψ be a G–equivariant symplectic diffeomorphism in V. Then there exists $x \in \Lambda$ with $\psi(x) \in L_\Lambda(x)$.

What happens if we drop the group action? As the following Theorem shows the situation is subtle and far from being understood.

THEOREM 13 Assume Λ is a compact connected smooth hypersurface of restricted contact type in V enclosing 0. Let $t \to \psi_t$ be an isotopy of the identity in D_ω fixing 0 and let $H: [0,1] \times V \to \mathbb{R}$ be the Hamiltonian generating $t \to \psi_t$ and normalized by $H(t,0) = 0$ for all $t \in [0,1]$. We assume

$$H(t,x) - \tfrac{1}{2}\langle H''(t,0)\, x,x\rangle \geq 0$$

for every $t \in [0,1]$ and x enclosed by Λ

and

$$\frac{1}{2}\int_0^1 \langle -J\dot{x},x\rangle dt - \int_0^1 H(t,x(t))dt \leq 0$$

for every fixed point x_0 of ψ_1 which is enclosed by Λ where $x(t) = \psi_t(x_0)$. Then there exists $x \in \Lambda$ satisfying $\psi_1(x) \in L_\Lambda(x)$.

If one relaxes the condition somewhat one can have nonexistence as an example in [E–H4] shows. Theorem 13 implies generalization of results due to F. Clarke [Cl1] and J. Moser [Mo1]. It is not known how good Theorem 13 is, i.e. how sharp are the hypotheses.

4. A RELATED SYMPLECTIC CAPACITY FUNCTION

We sketch now a proof of the existence of a particular symplectic capacity function as given in II.3 Theorem 7. Prompted by the proof of Lemma 4 in III. 1 we introduce a subgroup β of the homeomorphism group of $E = H^{\frac{1}{2}}(S'; V)$. Namely

$$(1) \quad h \in \beta \iff h: E \to E \text{ is a homeomorphism such that}$$
$$h(x) = e^{\gamma^+(x)}x^+ + x^0 + e^{\gamma^-(x)}x^- + K(x)$$

where $\gamma^+, \gamma^-: E \to \mathbb{R}$ are continuous and map bounded sets into bounded sets. Moreover K is continous and maps bounded sets into precompact sets. Further

$$(2) \quad K(x) = 0, \ \gamma^+(x) = \gamma^-(x) = 0 \ \text{ if } \ \|x\| \ \text{ large}$$
$$\text{or } \int_0^1 <-J\dot{x},x>dt \leq 0$$

That β is a subgroup of homeo (E) is not difficult to verify.

Given $\Lambda \in \ell_r$, i.e. Λ is of restricted contact type, we denote by $\mathcal{F}(\Lambda)$ the set of all $H: V \to [0,+\infty)$ such that

$$(3) \quad H(x) = 0 \ \text{ for all } x \text{ in an open neighborhood of } B_\Lambda \cup \Lambda.$$

and

$$(4) \quad H(x) = k|x|^2 \ \text{ for } |x| \text{ large for some } k \in (0,+\infty)$$

Finally we define for $\Lambda \in \ell_r$

$$(5) \quad \tau(\Lambda) = \inf_{H \in \ (\Lambda)} \ \sup_{h \in \beta} \ \inf_{x \in S^+} \phi_H(h(x))$$

where S^+ is the unitsphere in E^+ and ϕ_H is defined by

$$\phi_H(x) = \frac{1}{2}\left((-P^- + P^+)x,x\right) - \int_0^1 H(x(t))dt$$

Then we put

$$\alpha(B_\Lambda) = \tau(\Lambda) \quad \text{for} \quad \Lambda \in \ell$$

and for $U \neq \emptyset$ open

$$\alpha(U) = \sup \{\alpha(B_\Lambda) \mid B_\Lambda \subset U, \Lambda \in \ell_r\}$$

and finally for $S \in \varphi$

$$\alpha(S) = \inf \{\alpha(U) \mid U \supset S , U \text{ open}\}$$

One verifies then (which is quite technical) that α is a symplectic capacity function having the properties stated in Theorem 7.

IV. THE ARNOLD CONJECTURES

In the introduction we have already described some of the history of the Arnold conjectures. In the first section we describe the problems and introduce the necessary notation. We also suggest the reading of [Ar1].

1. SYMPLECTIC FIXED POINT THEORY AND LAGRANGIAN INTERSECTION THEORY

In the following we assume that (M,ω) is a compact symplectic manifold. We are interested in studying fixed points for symplectic diffeomorphisms on M. One cannot expect in general that every symplectic map has a fixed point. Just look at the map on $\mathbb{R}^{2n}/\mathbb{Z}^{2n}$ induced by $x \to x + (\frac{1}{2},0,0,...,0)$. However maps generated by a time dependent Hamiltonian vectorfield are good candidates. We say $\psi \in \mathcal{P}_{\omega}(M)$ is an exact symplectic diffeomorphism iff there exists a Hamiltonian H so that Ψ is the time-1 map for the flow associated to the Hamiltonian vectorfield $\dot{x} = X_{H_t}(x)$, $dH_t = \omega(X_{H_t},\cdot)$.

Arnold conjectured in [Ar1] certain lower bounds for the number of fixed points of an exaxt symplectic map. A more general concept than fixed point theory is Lagrangian intersection theory. A submanifold L of M is called Lagrangian if 2dim L = M and $\omega/L = 0$. Denote by $M \times M$ the symplectic manifold with symplectic form $\omega \oplus (-\omega)$. One easily verifies that the graph of a symplectic map ψ, say graph (ψ) is a compact Lagrangian submanifold, also diag$(M\times M)$, the diagonal is Lagrangian. Clearly we have a natural bijection

$$\mathrm{Fix}(\psi) \xrightarrow{\sim} \mathrm{graph}(\psi) \cap \mathrm{diag}(M\times M)$$

So the study of the fixed point problem is equivalent to the intersection problem. There are of course intersection problems which cannot be formulated as a fixed point problem.

The result whose proof we are going to sketch is the following, see [F7], [H3].

THEOREM 14 Let (M,ω) be a compact symplectic manifold. Assume L is a compact Lagrangian submanifold such that $\omega/\pi_2\,(M,L) = 0$. Assume L' is an exact deformation of L. Then $\#(L \cap L') \geq c(L)$, where $c(L)$ is the cohomological $\mathbb{Z}_2$–category which is defined below.

Here L' is an exact deformation of L iff $L' = \psi(L)$ for an exact symplectic diffeomorphism on M.

DEFINITION 7 The cohomological $\mathbb{Z}_2$–category is the smallest number k so that there exists an open covering U_1, U_k of L so that the inclusion maps $\epsilon_i\colon U_i \to (L,\{x_o\})$ for some fixed $x_o \in L$ induce the zero maps $\check{\epsilon}_i\colon \check{H}(L,\{x_o\})_i\ \mathbb{Z}_2) \to \check{H}(U_i)\ \mathbb{Z}_2)$ in Cech cohomology with coefficients in $\mathbb{Z}_2$.

A trivial consequence, as already sketched, is the following Theorem.

THEOREM 15 Let (M,ω) be a compact symplectic manifold such that $\omega|\pi_2(M) = 0$. Then an exact symplectic diffeomorphism on M has at least $c(M)$ fixed points.

It is possible to replace $c(M)$ by the cohomological category where we maximize the number k in the definition 7 over all commutative rings R which we take as coefficients. Such a generalization is not possible for Theorem 14.

We sketch now a proof of Theorem 14. We can construct an almost complex structure J on M, i.e. $J(x)\colon T_xM \to T_xM$. $J(x)^2 = -Id$ such that $g = \omega \circ (J \times I)$ is a Riemannian metric. We call such an almost complex structure positive (better ω–positive). Denote by Z the strip $\{s + it \mid s \in \mathbb{R}, t \in [0,1]\}$ in $\mathbb{C}$. We shall study the first order elliptic system on Z given by

$$(1)\quad u_s + J(u)u_t = 0\ , \quad u\colon Z \to M$$

$$u(\mathbb{R}) \subset L\ , \quad u(i + \mathbb{R}) \subset L'$$

$$\int_Z (|u_s|^2 + |u_t|^2) dsdt < \infty$$

A solution of (1) is automatically smooth by standard elliptic regularity theory. Denote by $\Omega = \Omega_J(L,L')$ the set of solutions of (1). We equip Ω with the compact–open C^∞–topology induced from $C^\infty(Z,M)$ and which is of course metrizable, say d is a metric. We have a continuous map $\Omega \xrightarrow{\pi} L$: $u \to u(o)$. The key step in proving Theorem 14 is the following

THEOREM 16 Under the hypothesis of Theorem 14 (Ω,d) is a compact metric space. Moreover $\check{\pi}$: $\check{H}(L) \to \check{H}(\Omega)$ is injective.

Next observe that we have a continous flow on Ω defined by

$$\mathbb{R} \times \Omega \to \Omega\text{: } (\tau,u) \to u\cdot\tau \quad , \ (u,t)(s+it) = u(s-\tau+it)$$

Moreover it is not difficult to show that there exists a strict Liapunov function α: $\Omega \to \mathbb{R}$, i.e. $\tau \to \alpha(u\cdot\tau)$ is strictly decreasing unless u is a fixed point for the flow. Clearly u can only be a fixed point for the flow if $u(s+it) = \text{const} \in L \cap L'$ since $\int |u_s|^2 + |u_t|^2 < \infty$. In fact one has

$$\alpha(u\cdot\tau) - \alpha(u) = -\frac{1}{2}\int_{[-\tau,o]+i[0,1]} (|u_s|^2 + |u_t|^2) dsdt$$

Summing up we have a gradient-like flow on a compact metric space. Now using methods as in [Co–Ze2] one sees easily that
Rest points of "·" $\geq c(L)$. Since Rest points of "·" $= L \cap L'$ the proof of Theorem 14 is complete.

2. GROMOV'S THEORY OF ALMOST HOLOMORPHIC CURVES

As we have seen the differential equation $u_s + J(u)\, u_t = 0$ plays an important role in the study of Lagrangian intersection problems. Note that this differential operator is basically a nonlinear Cauchy–Riemann operator. We shall therefore write

$$\bar{\partial}u := u_s + J(u)\, u_t$$

It was Gromov who pointed out in his seminal paper [Gr1] the importance of almost holomorphic curves in symplectic geometry. Gromov's method was non variational. Floer however was able to combine Gromov's ideas with the variational calculus, more precisely a homological Conley index.

There are two important analytical facts concerning certain maps $u: B(\epsilon) \to M$, where $B(\epsilon)$ is the ϵ–ball around 0 in $\mathbb{C}$. Using a theorem due to Nash we may assume that (M,g) is isometrically embedded in some large $(\mathbb{R}^N, <\cdot,\cdot>)$.

In other words we may assume that we are given a compact submanifold M of some $\mathbb{R}^N$. Moreover M is equipped with an almost complex structure J so that $\omega = <+J\cdot,\cdot>$ is a symplectic form on M.

Denote by $H^j(G,\mathbb{R}^N)$; $j = 0,1,2,$ the standard Sobolev spaces. By $H^j(G,M)$ we denote the subset consisting of these maps in $H^j(G,\mathbb{R}^N)$ which have their image in M. If $j \geq 2$ $H^j(G,\mathbb{R}^N)$ can be actually seen as a submanifold of the Hilbert space $H^j(G,\mathbb{R}^N)$ provided G is bounded. We give now two important facts.

FACT 1 Let (u_k) be a sequence of maps in $H^j(B(\epsilon), M)$, $j \geq 2$, such that $\|\bar{\partial}u_k\|_{j-1,B(\epsilon)} \to 0$ as $k \to \infty$. Assume $\|u_k\|_{j,B(\epsilon)} \leq c$ for some constant $c > 0$. There exists a $\delta \in (0,\epsilon)$ such that $(u_k \mid B(\delta))$ is precompact in $H^j(B(\delta), M)$.

Fact 1 is a consequence of basic linear elliptic estimates for $\bar{\partial}$.

The difficulty to apply Fact 1 is that one needs the a priori bound on $\|u_k\|_{j,B(\epsilon)}$. Here fact 2 is useful.

FACT 2 Let (u_k) be a sequence of maps in $H^j(B(\epsilon), M)$ such that $\|\bar\partial u_k\|_{j-1,B(\epsilon)} \to 0$ as $k \to \infty$. Assume $\|u_k\|_{1,B(\epsilon)} \leq c$ for some $c > 0$ and all k. If $\omega|\pi_2(M) = 0$ there exists $\delta \in (0,\epsilon)$ such that $(u_k|B(\delta))$ is precompact in $H^j(B(\delta), M)$.

Similarly one has the above estimates if $B(\epsilon)$ denotes the ball around 0 in $\mathbb{C}^+ = \{z \in \mathbb{C} \mid im(z) \geq 0\}$ and u_k maps the (boundary of $B(\epsilon)) \cap \partial\mathbb{C}^+$ into a compact Lagrangian submanifold, see [H3].

In fact 2 the assumption $\omega|\pi_2(M) = 0$ comes in as follows. If the a priori estimate does not hold we are able to use the conformal invariance of the operator $\bar\partial$ to construct a map $u: \mathbb{C} \to M$ $u \neq$ constant, $\int |u_s|^2 + |u_t|^2 < \infty$. One can apply a removable singularity theorem [P] to obtain an almost holomorphic map $\bar u: S^2 \to M$. Since $\bar u$ is nonconstant we have

$$\int_{S^2} \bar u^* \omega = \frac{1}{2} \int_{\mathbb{C}} |u_s|^2 + |u_t|^2 > 0$$

On the other hand $\omega|\pi_2(M) = 0$ so that $\int_{S^2} \bar u^* \omega = 0$ giving a contradiction. So we obtain a contradiction since bubbling off of a holomorphic sphere does not occur if $\omega|\pi_2(M) = 0$.

Note that Fact 1 and Fact 2 with the remark thereafter imply immediately the first part of Theorem 16 if one can show the existence of a constant $c_1 > 0$ such that every $u \in \Omega$ satifies

$$\frac{1}{2} \int_Z (|u_s|^2 + |u_t|^2) \leq C_1$$

or equivalently

$$\int_Z u^* \omega \leq C_1$$

This is however purely topological and follows from the calculus of differential forms, see [H3], using that L' is an exact deformation of L.

3. LUSTERNIK–SCHNIRELMAN THEORY AND ALMOST HOLOMORPHIC DISKS

In order to be able to obtain a Lusternik–Schnirelman–Theory we have to show that the map $\check{\pi}$ induces an injection in Čech Cohomology. In order to do so we use the continuity property of Čech cohomology together with an approximation result derived in [H3]. We choose a compact domain with smooth boundary ∂G in $\mathbb{C}$ such that $G \subset Z$ and $\partial G \cap \partial Z \supset [-1,1] + i\{0,1\}$. We also assume that G is convex. Now we stretch G to obtain domains G_φ for $\varphi > 0$ as follows. We cut G into two halfs G^-, G^+ along the axis $i\mathbb{R}$ and shift G^- to the left and G^+ to the right so that we can insert a piece $(-\varphi,\varphi) + i[0,1]$. Now using that L' is an exact deformation of L, say there is an exact isotopy ψ_t of the identity in $D_\omega(M)$ such that $\psi_1(L) = L'$ we define a map $\partial G_\varphi \to$ Lagrangian submanifolds of M by $x \to \psi_{\beta(x)}(L)$ having the following property. For simplicity write L_x instead of $\psi_{\beta(x)}(L)$

$$L_x = \begin{cases} L & x \in [-\varphi-1,\ \varphi+1] \\ L' & x \in [-\varphi-1,\ \varphi+1] + i \\ \text{interpolation between } L \text{ and } L' \text{ for the remaining } x. & \end{cases}$$

If we enlarge φ we can take the same interpolation between L and L' and just keep $L_x = L$ or $L_x = L'$ on a corresponding longer interval. We study now the elliptic boundary value problem

$$(P_\varphi) \quad \begin{array}{ll} \bar{\partial} u = 0 & \text{on } G_\varphi \\ u(x) \in L_x & \text{for } x \in \partial G_\varphi \end{array}$$

It is not difficult to see that we have a uniform H^1–bound for topological reasons independent of $\varphi > 0$ (just use some calculus with differential forms.) Using the estimates in section 2 one gets uniform

C^∞–bounds on every compact subset K of Z for the restriction of the solution of (P_φ) (φ large enough) to K. In view of the Ascoli–Arzela Theorem this is of course a strong compactness statement. Let $\gamma\colon \mathbb{R} \to [0,1]$ be a smooth map such that $\gamma(s) = 1$ for $s \le 1$ and $\gamma'(s) < 0$ for $1<s<2$ and $\gamma(s) = 0$ for $s \ge 2$. For u a solution of (P_φ) we define a new map $r_\varphi(u)\colon Z \to M$ by

$$r_\varphi(u)(s + it) = u(\gamma(\tfrac{2|s|}{\varphi})\, s + it)$$

provided φ is large enough. It is an easy exercise that r_φ defines a continuous map

$$\{u \in C^\infty(G_\varphi,M) \mid u(x) \in L_x\ x \in \partial G_\varphi\} \to \{u \in C^\infty(Z,M) \mid u(\mathbb{R}) \subset L,\ u(\mathbb{R}+i) \subset L'\}$$

Denote by Ω_φ the solution set of (P_φ). What we said about a priori estimates concerning (P_φ) immediately implies the following approximation result.

THEOREM 17 Under the hypothesis of Theorem 14 we find for a given open neighborhood U of $\Omega(L,L')$ in $\{u \in C^\infty(Z,M) \mid u(\mathbb{R}) \subset L,\ u(\mathbb{R} + i) \subset L'\}$ a number $\varphi_0 > 0$ such that for every $\varphi \ge \varphi_0$ we have $r_\varphi(\Omega_\varphi) \subset U$.

Define $\pi_U\colon U \to L$ by $u \to u(0)$. Assume now we can show for every $\varphi > 0$ that the map π_φ defined by

$$\pi_\varphi\colon \Omega_\varphi \to L\colon \quad u \to u(o)$$

induces an injective map in Čech cohomology. Then we see since

$$\check{r}_\varphi\, \check{\pi}_U = \check{\pi}_\varphi$$

that $\check{\pi}_U$ is injective as well. This implies since $\Omega(L,L')$ is a compact subset of a metric space by the continuity property of Čech cohomology that $\check{\pi}\colon \check{H}(L) \to \check{H}(\Omega(L,L'))$ is injective. In view of this remark we have to study

(P_φ) and show that $\pi_\varphi: \Omega_\varphi \to L$ induces an injection in cohomology. The advantage of (P_φ) in contrast to the elliptic problem defining $\Omega(L,L')$ is that (P_φ) is a problem on a bounded domain. We freeze now φ and consider the problem

$$(P) \quad \bar{\partial}u = 0 \text{ on } G_\varphi$$
$$u(x) \in L_x \text{ for } x \in \partial G_\varphi$$

We embed the problem (P) into a one parameter family of problems as follows: Recall that $L_x = \psi_{\beta(x)}(L)$ for a suitable map $\beta: \partial G_\varphi \to [0,1]$. Define for $\tau \in [0,1]$ a new family $x \to L_x^\tau$ by

$$L_x^\tau = \psi_{(\tau\beta(x))}(L)$$

Then $L_x^1 = L_x$ and $L_x^0 = L$ for every $x \in \partial G_\varphi$.

We consider now the family of nonlinear elliptic boundary value problems

$$(P^\tau) \quad \bar{\partial}u = o \text{ on } G_\varphi$$
$$u(x) \in L_x^\tau \text{ for } x \in \partial G_\varphi$$

For $\tau = 0$ we in fact know all the solutions, namely all constant maps with image in L. Moreover our a priori estimates imply that we can consider (P^τ) as a problem for a family of proper Fredholm operators. Define a Hilbert manifold Λ^j for $j \geq 2$ by

$$\Lambda^j = \{u \in H^j(G_\varphi, M) \mid u(\partial G_\varphi) \subset L\}$$

Moreover define for $\tau \in [0,1]$

$$\Lambda_\tau^j = \{u \in H^j(G_\varphi, M) \mid u(x) \in L_x^\tau \text{ for } x \in \partial G_\varphi\}.$$

We define a Hilbert space bundle $E^{j-1} \to H^j(G_\varphi, M)$ by

$$E_u^{j-1} = H^{j-1}(u^*TM)$$

where u^*TM is the pullback of the tangent bundle via $u: G_\varphi \to M$. By Kuipers theorem, [Ku], there exists a trivialization

$$E^{j-1} \xrightarrow{\Theta} H^j(G_\varphi, M) \times H$$

where H is an (abstract) separable Hilbert space. We can construct a smooth family of diffeomorphism of M, say $\phi_{\tau,x}: M \to M$, $x \in G_\varphi$ such that $\Gamma_\tau: \Lambda^j \to H^j(G_\varphi, M)$ defined by

$$\Gamma_\tau(u)(x) = \phi_{\tau,x}(u(x))$$

induces a diffeomorphism $\Lambda^j \longrightarrow \Lambda^j_\tau$ still denoted by Γ_τ. We consider now the 1-parameter family of smooth maps $f_\tau: \Lambda^j \to H$ defined by

$$f_\tau(u) = pr_2 \circ \Theta \circ \bar{\partial} \circ \Gamma_\tau(u)$$

where $pr_2: H^j(G_p, M) \times H \to H$ is the projection onto the second factor. It turns out that $f_\tau: \Lambda^j \to H$ is proper with respect to an open neighborhood of 0 in H. Hence degree theory is available. Moreover f_τ is a Fredholm operator of index n. We define a family of Fredholm operators $\tilde{f}_\tau: \Lambda^j \to H \times L$ by

$$\tilde{f}_\tau(u) = (f_\tau(u)\ ,\ \pi(u))$$

We see that $\tilde{f}_0^{-1}(0,\ell_0)$ for some fixed $\ell_0 \in L$ consists precisely of the constant solution $u_0(x) = \ell_0$. Moreover $T\tilde{f}_\tau(u_0): T_{u_0}\Lambda^j \to H \times T_{\ell_0}L$ is an isomorphism. Hence denoting the $\mathbb{Z}_2$-degree by $d_{\mathbb{Z}_2}$ we infer

$$d_{\mathbb{Z}_2}(\tilde{f}_0, (0,\ell_0)) = 1.$$

By homotopy invariance of the degree we obtain

$$d_{\mathbb{Z}_2}(\tilde{f}_1, (0,\ell_o) = 1.$$

Now the desired result concerning the behavior of π on cohomological level follows from the following abstract result, [H3].

THEOREM 18 Let $f: V \to H$ be a smooth Fredholm map of index n defined on a separable Hilbert manifold V with image in a separable Hilbert space H. Assume f is proper with respect to a zero neighborhood in H. Assume there exists a smooth map $\pi: V \to L$ into a compact manifold of dimension n such that

$$d_{\mathbb{Z}_2}(\tilde{f},(0,\ell_o) = 1$$

where $\tilde{f}(x) = (f(x), \pi(x))$ and $\ell_o \in L$ is fixed. Then

$$\check{\pi}: \check{H}(L) \to \check{H}(f^{-1}(o))$$

is injective.

For a simple proof see [H3].

4. MORSE THEORY AND FLOER HOMOLOGY

In 3 we gave a relatively simple proof of the Lusternik–Schnirelman theory for Lagrangian intersections. The Morse theory developed prior to the LS theory by A. Floer is very difficult. Actually the idea is quite simple but the technical realization is very hard. Here we shall sketch the simple idea. The idea was motivated by an influential paper of Witten [Wi]. The reader should read also [F5]. Consider a time dependent family of positive almost complex structures, say $t \to J_t$. We have an associated family of Riemannian metrics g_t on M.

We assume L and L' are compact Lagrangian submanifolds so that L' is an exact deformation of L and the intersections in $L \cap L'$ are transversal. We define for $J = (J_t)_{t\in[0,1]}$

$$\Omega_J = \{u \in C^\infty(Z,M) \mid u_s + J_t(u)u_t = 0 \ , \ \int |u_s|^2 + |u_t|^2 < \infty \\ u(\mathbb{R}) \subset L \ , \ u(\mathbb{R} + i) \subset L'\}$$

First one derives the same estimate we derived for the LS theory. Then one shows that for a generic family J Ω_J can be written as the union of finite manifolds $M(x,y)$ where $(x,y) \in (L\cap L')^2$

$$\Omega_J = \cup M(x,y)$$

where $M(x,y)$ consists of those $u \in \Omega_J$ such that $u \to x,y$ as $s \to \pm\infty$. Moreover there exists a unique map μ: $L \cap L' \to K$ defined up to an additive constant such that

$$\dim M(x,y) = \mu(x) - \mu(y)$$

for generic J. Moreover we have an obvious $\mathbb{R}$–action on $M(x,y)$. Denote by $\hat{M}(x,y) = M(x,y)/\mathbb{R}$ the reduced space. Clearly

$$\dim \hat{M}(x,y) = \mu(x) - \mu(y) - 1$$

It turns out now that under the assumption $\pi_2(M,L) = 0$, the set $\hat{M}(x,y)$ is finite if $\mu(y) - \mu(x) = 1$ (for generic J).

Denote by C the free $\mathbb{Z}_2$–vector space generated by the points in $L \cap L'$. C is a graded vector space where the grading is given by the map μ, i.e.

$$C^P = \text{span}\ \{x \mid \mu(x) = p\}.$$

So $C = \oplus C^P$. We define a coboundary operator δ: $C \to C$ of degree 1 by

$$\delta x = \sum_{\mu(y)=\mu(x)+1} < y,\delta x> y$$

where $<y,\delta x>$ is the number of elements in $\hat{M}(y,x)$ mod 2. (recall that $\hat{M}(y,x)$ is finite if $\mu(y) = \mu(x) + 1$. A crucial observation is then $\delta^2 = 0$. So we can define the cohomology of the complex (C,δ)

$$I^*(L,L') = \mathrm{kern}(\delta)/\mathrm{im}(\delta)$$

$I^*(L,L')$ is called the Floer Cohomology of the intersection problem $(M,(L,L'))$.

Assume now L_τ is an exact homotopy of $L = L_0$. In [F2] it is shown that

(1) $I^*(L,L')$ does not depend on the choice of J as long as J is generic.

(2) $I^*(L,L_{\tau_1}) \simeq I^*(L,L_\tau)$ provided L and L_{τ_1}, L_τ: intersect transversally.

(3) $I^*(L,L_\tau) \simeq H^*(L,\mathbb{Z}_2)$.

Clearly if we have only transversal intersections in $L \cap L'$ and we associate to an intersection point x the monomial $t^{\mu(x)+c}$ for some fixed integer c (which we assume to be zero), we obtain the equality

$$\sum_{x \in L \cap L'} t^{\mu(x)} = P(t) + (1+t)Q(t)$$

where P(t) is the $\mathbb{Z}_2$–Poincare polynomial of L. So we have the following

THEOREM 19 If (M,ω) is a compact symplectic manifold, and L a compact Lagrangian submanifold satisfying $\omega | \pi_2(M,L)$ and L' an exact deformation of L intersecting L' transversally then there exists a unique map $\mu: L \cap L' \to \mathbb{Z}$ such that

$$\sum_{x \in L \cap L'} t^{\mu(x)} = P(t) + (1+t)Q(t)$$

where P is the Poincaré' Polynomial of L associated to H with $\mathbb{Z}_2$-coefficients and $Q(t)$ is a polynomial with nonnegative coefficients in $\mathbb{Z}$.

If $\pi_2(M,L) \neq 0$ one can give easily counterexamples for the Lagrangian intersections problem. However one expects some results for the fixed point problem if $\pi_2(M) \neq 0$. The first such result was given in [Fo] by B. Fortune. Recently in [F6] Floer proved some further results. The difficulty is that one does not have in general compactness and the phenomenon of bubbling off of holomorphic spheres can occur and has to be taken into account.

V. REFERENCES

[Ab–Ma] R. Abraham – J. Marsden, Foundation of Mechanics Benjamin/Cummings Reading, Man., 1978.

[Ag–Do–Ni] S. Agmon – A. Douglis – L. Nirenberg, Estimates near the boundary for solutions of elliptic partial differential equations satisfying general boundary conditions I, II Comm. Pure Appl. Math. 12(1959) p. 623–727, 17(1964), 35–92.

[A–M] A. Ambrosetti – G. Mancini, Solutions of minimal period for a class of convex Hamiltonian systems, Math. Ann. 255(1981), 405–421.

[Am–Ze] H. Amann – E. Zehnder, Nontrivial solutions for a class of nonresonance problems and applications to nonlinear differential equations, Ann. Scu. Norm. Sup. di Pisa 7(1980), 593–603.

[Ar1] V.I. Arnold, Sur une propriete topologique des applications canoniques de la mecanique classique, C.R. ASP t. 261(1965), 3719–3722.

[Ar2] V.I. Arnold, Mathematical methods of classical mechanics, Springer.

[Aro] N. Aronszajn, A unique continuation theorem for solutions of elliptic partial differential equations or inequalities of second order, J. Math. Pures et Appl. (9) 36(1957), 235–249.

[AS1] M.F. Atiyah – I.M. Singer, The index of elliptic operators I, III–V; I: Ann. Math. 87(1968) 484–530 III: Ibid 87(1968) 546–604, IV: Ibid 93(1971) 119–138, V: ibid 93(1971), 139–149.

[AS2] M.F. Atiyah – I.M. Singer, Index theory of skew adjoint Fredholm operators, Publ. Math. IHES 37(1969), 305–325.

[Ba] A. Banyaga, Sur la groupe des diffeomorphismes qui preservent une forme symplectique, Comm. Math. Helv 53(1978), 174–227.

[Be–Ra1] V. Benci – P.H. Rabinowitz, A priori bounds for periodic solutions of Hamiltonian systems, to appear in Ergodic theory and dynamical systems.

[Be–Ra2] V. Benci – P.H. Rabinowitz, Critical point theorems for indefinite functionals, Inv. Math. 52(1979), 241–273.

[Be–Ho–Ra] V. Benci– H. Hofer – P.H. Rabinowitz, A remark on a priori bounds and existence for periodic solutions of Hamiltonian systems, Nato ASI Series C Vol. 209, 85–89.

[B] D. Bennequin, Problems elliptiques, surfaces de Riemann et structures symplectique, Sem. Bourlaki 38(1985–86), No. 657.

[Bor–Z–S] V. Borisovich – V. Zvyagin – V. Sapronov, Nonlinear Fredholm maps and Leray–Schauder–Degree, Russian Math. Survey's 32:4(1977), 1–54.

[Ch] M. Chaperon, Quelques questions de geometrie symplectique, Sem. Bourbaki 1982/1983, Asterisque 105–106, 231–249.

[Cl1] F. Clarke, On Hamiltonian Flows and symplectic transformations, Siam J. of control and Optimization, vol 20(3) (1982), 355–359.

[Cl2] F. Clarke, Periodic solutions of Hamiltonian inclusions, J. Diff. Eq. 40(1981), 1–6.

[Cl–E] F. Clarke – I. Ekeland, Hamiltonian trajectories having prescribed minimal period, Comm. Pure and Appl. Math. 33(1980), 103–116.

[Co] C.C. Conley, Isolated invariant sets and the Morse index, CBMS Reg. Conf. Series in Math 38, AMS 1978.

[Co–Ze1] C.C. Conley – E. Zehnder, the Birkhoff–Lewis fixed point theorem and a conjecture by V.I. Arnold, Inv. Math 73(1983), 33–49.

[Co–Ze2] C.C. Conley – E. Zehnder, Morse type index theory for flows and periodic solutions for Hamiltonian equations, Comm. Pure Appl. Math. Vol. 27(1984), 211–253.

[E1] I. Ekeland, Une theorie de Morse pour les systemes hamiltoniens convexes, Ann. IHP Analyse non lineaire (1984), 19–78.

[E2] I. Ekeland, An index theory for periodic solutions of convex Hamiltonian systems Proc. AMS summer institute on Nonlinear Functional Analysis 1983, Berkeley.

[E–H1] I. Ekeland – H. Hofer, Periodic solutions with prescribed minimal period for convex autonomous hamiltonian systems, Inv. Math. 81(1985), 155–188.

[E–H2] I. Ekeland – H. Hofer, Subharmonics for convex nonautonomous Hamiltonian systems, Comm. Pure Appl. Math. vol XL(1987), 1–36.

[E–H3] I. Ekeland – H. Hofer, Convex Hamiltonian energy surfaces and their periodic trajectories, to appear CMP.

[E–H4] I. Ekeland – H. Hofer, Two symplectic fixed point theorems, to appear.

[E–H5] I. Ekeland – H. Hofer, Symplectic Topolgy and Hamiltonian systems I, to appear.

[E–L] I. Ekeland – J.M. Lasry, On the number of periodic trajectories for a Hamiltonian flow on a convex energy surface, Ann. of Math. (2) 112(1980), 283–319.

[E11] H. Elliason, Geometry of manifold of maps, J. Diff. Geom. 1(1967), 165–194.

[F1] A. Floer, A proof of the Arnold conjecture for surfaces and generalizations to certain Kahler manifolds, Duke–Math J. 53(1986), 1–32.

[F2] A. Floer, Morse theory for Lagrangian intersections, to appear J. Diff. Geom.

[F3] A. Floer, The unregularized gradient flow of the symplectic action, to appear Comm. Pure Appl. Math.

[F4] A. Floer, A relative Morse index for the symplectic action, to appear Comm. Pure Appl Math.

[F5] A. Floer, Witten's complex for general coefficients and an application to Lagrangian intersections, to appear.

[F6] A. Floer, The Conley Index for the symplectic action, preprint (preliminary version).

[F7] A. Floer, Cup length estimates on Lagrangian intersections, to appear.

[F–H–V] A. Floer – H. Hofer – C. Viterbo, the Weinstein conjecture in $P \times \mathbb{C}^{\ell}$, to appear.

[Fo] B. Fortune, A symplectic fixed point theorem in $\mathbb{C}P^n$, Inv. Math. 81(1985), 29–45.

[Gr1] M. Gromov, Pseudo–holomorphic curves in symplectic manifolds, Inv. Math. 82(1985), 307–347.

[Gr2] M. Gromov, Partial differential relations, Springer 1986.

[Gr3] M. Gromov, Symplectic geometry hard and soft ICM Talk, Berkeley 1986, to appear.

[H1] H. Hofer, On strongly indefinite functionals with applications, TAMS 275(11) (1983), 185–214.

[H2] H. Hofer, Lagrangian embeddings and critical point theory, Ann. IHP Vol. 2 No.6(1986), 407–462.

[H3] H. Hofer, On the $\mathbb{Z}_2$–cohomology of certain families of almost holomorphic disks in symplectic manifolds and Lusternik–Schnirelman–Theory for Lagrangian intersections, to appear Ann. IHP.

[H–Vi] H. Hofer – C. Viterbo, The Weinstein conjecture in cotangent bundles and related topics, to appear.

[H–Ze] H. Hofer– E. Zehnder, Periodic solutions on hypersurfaces and a theorem by C. Viterbo, Inv. Math. 90(1987), 1–9.

[Ho] L. Hormander, The Analysis of linear differential operators III, Springer 1985.

[KL1] W. Klingenberg, Riemannian Geometry, de Gruyter studies in Math. 1, 1982

[KL2] W. Klingenberg, Lectures on closed geodesics, Springer 1978.

[Ku] N. Kuiper, The homotopy type of the unitary groups of Hilbertspace, Topology 3(1965), 19–30.

[L–S] F. Laudenbach – J.C. Sikorav, Persistence d'intersection avec la section nulle au cours d'une isotopie hamiltonienne dans un fibre cotangent, Inv. Math. 82(1985), 349–357.

[Lo–Mc] R.B. Lockhard – R.C. McOwne, Elliptic operators on noncompact manifolds, Ann. Sc. Norm. Sup. Pisa IV 12(1985), 409–446.

[McD] D. McDuff, Examples of symplectic structures, Inv. Math. 89(1987), 13–36.

[Mi] J. Milnor, Lectures on the H–cobordism theorem, Princeton University Press, Princeton 1965.

[Mo1] J. Moser, A fixed point problem in symplectic geometry, Acta. Math. 141(1–2) (1978), 17–34.

[Mo2] J. Moser, Periodic solutions near equilbrium and a theorem by A. Weinstein, Comm. Pure Appl. Math. 29(1976), 727–747.

[Pa1] R. Palais, Foundations of global nonlinear analysis, Benjamin, 1968.

[Pa2] R. Palais, Morse–theory on Hilbert manifolds, Topology 2(1963), 299–340.

[P] P. Pansu, Sur l'article de M. Gromov, Preprint.

[Ra1] P.H. Rabinowitz, Periodic solutions of Hamiltonian systems, Comm. Pure Appl. Math. 31(1987), 157–184.

[Ra2] P.H. Rabinowitz, Periodic solutions of a Hamiltonian system on a prescribed energy surface, J. Diff. Eq. 33(1979), 336–352.

[Ra3] P.H. Rabinowitz, Periodic solutions of Hamiltonian systems: a survey, Siam Review Math. Analysis Vol. 13(No. 3) (1982), p. 343–352.

[Ra–Am–Ek–Ze] Periodic Solutions of Hamiltonian systems and related topics, Ed. P.H. Rabinowitz, A. Ambrosetti, I. Ekeland, E.J. Zehnder, Nato ASI Series C Vol. 209.

[S–U] J. Sacks – K. Uhlenbeck, The existence of minimal immersions of two spheres, Ann. Math. 113(1981), 1–24.

[Si] J.C. Sikorav, Points fixes d'un symplectomorphisme homologue a l'identite, J. Diff. Geom. 22(1986), 49–79.

[Sim] C.P. Simon, A bound for the fixed point index of an area preserving map with applications to mechanics, Inv. Math. 26(1974), 187–200.

[Sm] S. Smale, An infinite dimensional version of Sard's theorem, Ann. J. Math. 87(1965), 861–866.

[Sp] E. Spanier, Algebraic topology, McGraw–Hill, NY, 1966.

[T] F. Takens, Hamiltonian systems: Generic properties of closed orbits and local perturbations, Math. Ann. 188(1970), 304–312.

[Ve] I.N. Vekua, Generalized analytic functions, Pergamon, 1962.

[Vi] C. Viterbo, A proof of the Weinstein conjecture in $\mathbb{R}^{2n}$, Ann. IHP analyse nonlineaire, (1987).

[Wa] W. Warschawski, On differentiability at the boundary in conformal mapping, Proc. AMS 12(1961), 614–620.

[We1] A. Weinstein, Lagrangian submanifolds and Hamiltonian systems, Ann. Math. 98(1973), 337–410.

[We2] A. Weinstein, C^0–Perturbation theorem for symplectic fixed points and Lagrangian intersections, Seminaire Sud–Rhodanien de geometrie III, Travaux on cours Herman Paris (1984), 140–144.

[We3] A. Weinstein, On the hypothesis of Rabinowitz's Periodic orbit theorems, J. Diff. Eq 33(1979), 353–358.

[We4] A. Weinstein, Symplectic Geometry and the Calculus of variations, Marston Morse Memorial Lectures, IAS Princeton, 1985, to appear.

[We5] A. Weinstein, Periodic orbits for convex Hamiltonian systems, Ann. Math. 108(1978), 507–518.

[Wi] E. Witten, Supersymmetry and Morse theory, J. Diff. Geom. 17(1982), 661–692.

[Wo] J.G. Wolfson, A. PDE proof of Gromov's compactness of pseudoholomorphic curves, preprint.

[Wen] W.L. Wendland, Elliptic systems in the plane, Pitman, 1979.

[Ze1] E. Zehnder, Periodic solutions of Hamiltonian systems, Lec. Notes in Math. 1031(1983), 1172–1213.

[Ze2] E. Zehnder, Remarks on periodic solutions on hypersurfaces, Nato ASI Series C Vol. 209, 267–280.

Solvability of Semilinear Operator Equations and Applications to Semilinear Hyperbolic Equations

P.S. MILOJEVIĆ Department of Mathematics, New Jersey Institute of Technology, Newark, New Jersey 07102

I. INTRODUCTION

Let H be a separable real Hilbert space, X be a Banach space densely and continuously embedded in $H, A : D(A) \subset H \to H$ be a densely defined closed linear map and N be a nonlinear map from $D(A) \cap X$ into H. We are interested in studying operator equations of the form

$$Ax - Nx = f \tag{1.1}$$

where $f \in H$ is given.

Equations of this form appear in a variety of situations, and in particular in the theory of ordinary and partial differential equations. For example, they can describe nonlinear elliptic boundary value problems, or problems concerning periodic solutions of semilinear hyperbolic equations, or Hamiltonian systems of ordinary differential equations, etc.

Eq. (1.1) has been studied extensively by various topological as well as variational methods. When A is a Fredholm map of nonnegative index, depending on the nature of a nonlinearity, various degree theories (e.g., for the compact or condensing perturbations of the identity, coincidence degree, etc.) have been used in conjunction with the Liapunov–Schmidt technique (cf. the references).

When the null space of A is infinite dimensional, Eq. (1.1) is much harder to study and the most often used topological approach is a combination of the Leray–Schauder and coincidence degrees and the monotone operator theory (cf. Brezis–Nirenberg [Br - Ni - 1–2], Mawhin [Ma -2 -4] and the references in there). Essential to this approach is the existence of a compact partial inverse $A^{-1} : R(A) \to H$ *of* A, where the range $R(A)$ of A is closed. However, if A contains also nonzero eigenvalues of infinite multiplicity, then A^{-1} is not compact and this approach is not suitable. This is the situation that occurs when studying the existence of periodic weak solutions of semilinear wave equations in more than one space variable. We note that, using the monotone operator theory and certain

This research was partially supported by the SBR Grant, NJIT approximation procedure, Amann [Am–2] was able to obtain rather general unique solvability results for Eq. (1.1) without requiring the compactness of A^{-1}.

In [Mi–5], we initiated a new approach for the study of Eq. (1.1) (with $\dim\ker A = \infty$) based on a Galerkin type method. This approach requires that $A + N$ is pseudo A–proper w.r.t. a scheme $\Gamma = \{H_n, H_n, P_n\}$ for (X,H), i.e. that the corresponding Galerkin procedure leads to a solution of (1.1). Here $\{H_n\}$ is a sequence of finite dimensional subspaces of $D(A) \cap X$, whose union is dense in both X and H, and $P_n : H \to H_n$ are the orthogonal projections. The only topological tool the method requires is the Brouwer degree theory and, as shown in [Mi–5–10], it is applicable to the situations studied by the above authors as well as to many new ones when neither A^{-1} is compact nor N is of monotone type.

In this paper we shall present a solvability theory for Eq. (1.1) using the pseudo A–proper mapping approach and give applications to the problem of existence of periodic weak solutions of (systems of) hyperbolic equations in one and more space variables. Both nonresonance as well as resonance cases will be considered, i.e. when the nonlinearity N stays away from the spectrum $\sigma(A)$ of A, or interacts with it in some way.

Part II of the paper is devoted to nonresonance problems for (1.1). We begin by assuming, roughly speaking, that N is asymptotically close at infinity to a selfadjoint map N_∞ and that

$$0 \notin \sigma(A - N_\infty).$$

This is a kind of nonresonance condition at infinity. Assuming that $(A - N_\infty)^{-1} : X \to H$ is continuous and $A - N$ is pseudo A–proper from X into H, we deduce the solvability of $Ax - Nx = f$ in X for each $f \in H$ (Theorem 2.1). An extension to a nonselfadjoint A and still another nonresonant result suitable for studying telegraph equations with nonlinear dissipation, are also given in Section 2.1. We note that when X is compactly embedded in H, then $A - N : D(A) \cap X \subset X \to H$ is pseudo A–proper without any monotonicity assumptions on N.

In Section 2.2, we continue our study of Eq. (1.1) without resonance involving the so-called asymptotically $\{B_1, B_2\}$–quasilinear nonlinearities, where B_1 and B_2 are given selfadjoint maps in H with $B_1 \leq B_2$. Such maps have been introduced in a more special case by Perov [Per] and Krasnoselskii–Zabreikko $[Kr - Za]$ in their study of the existence of fixed points of compact maps. The

additional condition that $\{B_1, B_2\}$ form a regular pair assumed by them is not needed in our study. The class of such maps is rather large. For example, if N has a symmetric weak Gateaux derivative $N'(x)$ and $B_1 \le N'(x) \le B_2$ for each $x \in H$, then N is asymptotically $\{B_1, B_2\}$-quasilinear.
Assume that there are closed subspaces $H^{\pm}$ of H such that $H = H^- \oplus H^+$ and that for some positive constants γ_1, and γ_2 and selfadjoint maps $B^{\pm}$ with $B_1 \le B^- \le B^+ \le B_2$

$$((A - B^-)x, x) \le -\gamma_1 \; ||x||^2 \text{ for } x \in D(A) \cap H^- \tag{1.2}$$

$$((A - B^-)x, x) \ge \gamma_2 \; ||x||^2 \text{ for } x \in D(A) \cap H^+. \tag{1.3}$$

We show (Theorem 2.4) that Eq. (1.1) is solvable for each $f \in H$ if, in addition, for some selfadjoint map C_0 with $B_1 - \epsilon I \le C_0 \le B_2 + \epsilon I$ for some $\epsilon > 0, H_t = A - (1-t)C_0 - tN$ is A-proper in H for each $t \epsilon [0, 1)$ and H_0 is pseudo A-proper.

In particular, this is so if, e.g., N is monotone and A^{-1} is compact, or $N = N_1 + N_2$ with N_1 strongly monotone and k_1-contractive and N_2 k_1-contractive with k_1, k_2 sufficiently small, and A^{-1} is just continuous (see Section 2.2 for a detailed study).

Conditions (1.2) and (1.3) are of nonresonance type and are satisfied if, for example, $B^- \le N'(x) \le B^+ \; on \; H$ and

$$\bigcup_{i=1}^{m} [\lambda_i^-, \lambda_i^+] \subset \varrho(A), \tag{1.4}$$

where $\varrho(A)$ is the resolvent set of A and $\lambda_i^{\pm}$ are pairwise disjoint eigenvalues of $B^{\pm}$ (cf. Amann [Am–2]). Section 2.2 contains even more general results (Theorems 2.5 and 2.6), which do not require that $H = H^- \oplus H^+$, as well as detailed applications of these results to various classes of A and N. In particular, the basic results of Amann [Am–2], Theorems 2.6 and 2.10, are extended to more general classes of nonlinearities. The proofs of the above results are based on some continuation results of the author [Mi –3]for (pseudo) A-proper maps.

The final section of Part II deals with the unique solvability of Eq. (1.1) and the convergence and error estimates of the approximate solutions. For example, if N is Gateaux differentiable and

$$a \, \| x - u \|^2 \le (Nx - Ny, x - y) \le \beta \, \| x - y \|^2 \quad , x, y \in H \tag{1.5}$$

with $[\alpha, \beta] \subset \varrho(A)$, then Eq. (1.1) is uniquely approximation solvable for each $f \in H$ and the approximate solutions satisfy

$$\| x_n - x_0 \| \leq c \| x_0 - P_n x_0 \| \leq c_1 \ dist(x_0, H_n).$$

This result is a constructive extension of a well-known uniqueness result of Amann [Am -1].

Part III deals with applications of the abstract theory developed in Part II to the problem of the existence of weak T-periodic solutions of various classes of semi-linear hyperbolic equations without resonance of the form

$$(1.6) \qquad \begin{cases} u_{tt} - L_1 u + \sigma u_t - F(t,x,u) = f(t,x), & t \in R, x \in Q \\ u(t,x) = 0 \ \text{ for } \ t \in R, x \in \partial Q, \end{cases}$$

where Q is a bounded domain in R^n with smooth boundary, $F: RxQxR^m \to R^m$ is a Caratheodory function, L_1 is a linear selfadjoint elliptic operator in space variables $x \in R^n$ with coefficients independent of t, $f \in L_2((0,T)xQ, R^m)$ with $\tau = 2\pi/T$ rational and $\sigma \in R$.

In Section 3.2, we study the semilinear wave equation (1.6), $\sigma = 0$, under various nonresonance conditions. We prove first the existence of T-periodic weak solutions in $\overset{\circ}{W}{}_2^1(\Omega, R^m)$, $\Omega = (0,T)xQ$, for each $f \in L_2(\Omega, R^m)$ assuming that F has a linear growth and, for example, for some consecutive eigenvalues $\lambda_i < \lambda_{i+1}$ of the associated linear problem and $1 \leq l \leq m$:

$$\lambda_i + \epsilon \leq F_l(t,x,y_1,\ldots,y_l,\ldots,y_m)/y_l \leq \lambda_{i+1} - \epsilon \tag{1.7}$$

for $a.e.(t,x) \in \Omega$, some $\epsilon > 0$ and all $y = (y_1,\ldots,y_l,\ldots,y_m) \in R^m$ *with* $|y_l| \geq R$. No monotonicity on F is required. We note that when $n = m = 1$ and $L_1 u = u_{xx}$, the solvability (1.5) with $\sigma = 0$ for a dense set of f's in L_2 was established earlier by Hofer [Ho-1] assuming also that f is a Lipschitz function (cf. also Wilhem [W]) and by Tanaka [Ta -1] without this additional condition. Next, assuming monotonicity on F, condition (1.7) is relaxed to the nonuniform one of the form in (Theorem 3.3):

$$\lambda_i \le \alpha_l(t,x) \le F_l(t,x,y_1,\dots,y_l,\dots,y_m)/y_l \le \beta_l(t,x) \le \lambda_{i+1} \tag{1.8}$$

for some $\alpha_l, \beta_l \in L_\infty(\Omega)$ with $\lambda_i < \alpha_l(t,x)$ and $\beta_l(t,x) < \lambda_{i+1}$ on some subset of Ω of positive measure. When $n = m = 1$ and $L_1u = u_{xx}$, this result is due to Mawhin - Ward [Ma–Wa] and was proved by them using rather different techniques.

The last application in Section 3.2 (Theorem 3.4) deals with a nonresonance condition of the form (1.4), where $\lambda_i^\pm$ are eigenvalues of symmetric matrices $b^\pm \in \mathcal{L}_s(R^m)$ and $F = F_1 + F_2$ is such that F_2 is k_2-contractive and $F_1(t,x,.) \in C^1(R^m, R^m)$, has a symmetric derivative $D_yF_2(t,x,y) \in \mathcal{L}_s(R^m)$ for $a.e.(t,x)$ and *all* $y \in R^m$ and satisfies

$$b^- \le D_yF_1(t,x,y) \le b^+ \ \text{ for } a.e.(t,x) \in R^m. \tag{1.9}$$

Condition (1.4) means that the eigenvalues of a symmetric $m \times m$ matrix $D_yF_1(t,x,y)$ lie in possibly distinct gaps of the spectrum of the abstract realization A of the linear problem. When $F_2 = 0$, this result is due to Amann [Am - 2], and in various special cases to Lazer [La], Ahmad [A], Brown–Lin [Br - Li] and Mawhin [Ma - 3]. Their techniques of proof do not apply to our situation.

Existence of *T*-periodic weak solutions for semilinear telegraph equation (1.6), $\sigma \ne 0$, is studied in Section 3.3 under various nonresonance conditions (like, e.g., of the form (1.7)) and without any monotonicity assumptions of F. Moreover, we have also proved the existence of *T*–periodic weak solutions in $\overset{\circ}{W}{}_2^1(\Omega, R^m)$ for a more general problem without resonance:

$$\text{(1.10)} \qquad \begin{cases} u_{tt} + L_1u + F(t,x,u_t) - G(t,x,u) = f(t,x) \ \text{ in } \Omega \\ u(t,x) = 0 \ \text{ on } [0,T]x\partial Q, \end{cases}$$

where again no monotonicity of F or G is required. These results (Theorems 3.6 and 3.7) are vast extensions of the corresponding ones of Prodi [Pr], who required a strong monotonicity and contractivity conditions on F.

Resonance problems for Eq. (1.1) are studied in Part IV. In the first part of Section 4.1, we established the solvability of (1.1), without monotonicity of N, under resonance conditions of the antipodes type and of Landesman–Lazer type. In the second part, we extend the basic result of Brezis–Nirenberg [Br– Ni – 1] to pseudo A–proper maps $A + N$. More precisely, let $a > 0$ be the largest number such that

$$(Ax, x) \geq -a^{-1} \| Ax \|^2 \quad on \ D(A), \qquad R(A) = N(A)^{\perp}$$

and suppose that for some $\gamma < a$ the following resonance condition holds:

$$(Nx - Ny, x) \geq \gamma^{-1} \| Nx \|^2 - c(y) \quad \text{for} \quad x, y \in H. \tag{1.11}$$

Assuming that $A - N : D(A) \subset H \to H$ is pseudo A–proper, we show that

$$Int(R(A) + convR(N)) \subset R(A - N)$$

and $A - N$ is onto, if N is also. Here, *Int(D)* and *conv D* denote the interior and the convex hull of D, respectively. When A^{-1} is compact and N is monotone, we deduce the above mentioned result in [Br – Ni – 1]. If A^{-1} is not compact, the above result is also applicable to nonlinearities satisfying (1.2) – (1.3) with $\gamma_1 = \gamma_2 = 0$ and (1.4) with $\varrho(A)$ replaced by its closure – which then is a resonance condition.

Section 4.2 is devoted to the study of (1.1) via a perturbation method. It consists in finding conditions on A, N and f which imply that, for some bounded map G, the perturbed equations

$$Ax + Nx + \epsilon Gx = f \tag{1.12}$$

have a solution x_ϵ for each $\epsilon > 0$ small, $\{x_\epsilon\}$ is bounded as $\epsilon \to 0$ and its weak limit is a solution of (1.1). Assuming that $A + N$ has a certain closedness property, that (1.12) is solvable for each $\epsilon > 0$ small and that $f = f^* + f^{**}$ with $f^* \in R(A)$ and for some $\gamma < a$ and c:

$$(Nx - f^{**}, x) \geq \gamma^{-1} \| Nx \|^2 - c \ \text{ for } \ x \in H,$$

we show that Eq. (1.1) is solvable if, for example, N is coercive on the *ker A*. In particular, if $N = \partial\psi$ for some convex function $\psi : H \to R$ and

$$\limsup_{\| x \| \to \infty} \| Nx \| / ||x|| < a/2,$$

then Eq. (1.1) is solvable if and only if $f = f^{*} + f^{**}$ with $f^{*} \in R(A)$ and $f^{**} \in R(N)$, provided additionally that either A^{-1} is compact, or $\sigma(A) \cap (0, \infty) \neq \phi$ and consists of eigenvalues of finite multiplicities, or conditions (1.2) – (1.3) hold with $\gamma_1 = \gamma_2 = 0$ and $B^- \leq N'(x) \leq B^+$.

Applications of the abstract results to resonance problems for hyperbolic equations of the form (1.6), $\sigma = 0$, are given in Part V. Suppose that $F = \partial G$ for some $G : \Omega x R^m \to R$ continuous and convex in $y \in R^m$. We show that (1.6), $\sigma = 0$, has a T-periodic weak solution for each $f \in L_2$ if $F = F_1 + F_2$ is such that F_1 is strongly monotone and k_1-contractive in $y \in R^m$, F_2 is k_1-contractive in y with k_1 and k_2 sufficiently small and for some $a < a$

$$|F(t, x, y)| \leq a|y| + h(t, x) \ \ a.e. \ on \ \Omega, y \in R^m \tag{1.13}$$

for some $h \in L_2$. On the other hand, if G is also $C1$ in y and

$$\int_{\Omega} G(t, x, u) dtdx \to \infty \ \ as \ \ ||u||_{L_2} \to \infty, u \in N(A)$$

and (1.13) holds, then there is a T-periodic weak solution of (1.6), $\sigma = 0$, for a given $f \in L_2$ if $f = f^{*} + f^{**}$ with $f^{*} \in R(A)$ and

$$(F(t, x, y) - f^{**}(t, x)) \cdot y \geq c|F(t, x, y) - f^{**}(t, x) \| y|$$

for all $|y| \geq R, a.e. (t, x) \in \Omega$ and some positive c and R.

II. SEMILINEAR EQUATIONS WITHOUT RESONANCE

In this part we shall study nonresonance problems for Eq. (1.1) assuming that A–N is a pseudo A-proper map. We begin by studying such problems involving not necessarily monotone nonlinear pertubations of both selfadjoint and nonselfadjoint linear maps with possibly infinite dimensional null space. We continue our study of Eq. (1.1) involving asymptotically $\{B_1, B_2\}$-quasilinear nonlinearities under the nonresonance conditions introduced by Amann [Am-2]. Finally, the unique approximation-solvability of Eq. (1.1) and the rate of convergence of the approximate solutions is proved under some stronger nonresonance conditions.

2.1 Semilinear Equations with Nonmonotone Nonlinearities

We begin by defining precisely the class of (*pseudo*) *A*-proper maps. Let X and Y be separable Banach spaces, $\{X_n\}$ and $\{Y_n\}$ be finite dimensional subspaces of X and Y with $\dim X_n = \dim Y_n$ and *dist* $(x, X_n) \to 0$ as $n \to \infty$ for each $x \in X$. If $Q_n : Y \to Y_n$ are linear projections such that $\delta = \max \| Q_n \| < \infty$, then $\{X_n, Y_n, Q_n\}$ is a projection scheme for (X,Y).

Let $D \subset X$ and $T: D \to Y$. Recall [Pet]

Definition 2.1 A map $T: D \subset X \to Y$ is *A*-proper (resp., pseudo *A*-proper) w.r.t.Γ if $Q_n T: D \cap X_n \to Y_n$ is continuous for each n and, whenever $\{x_{n_k} \mid x_{n_k} \in D \cap X_{n_k}\}$ is bounded and $Q_{n_k} T x_{n_k} - Q_{n_k} f \to 0$ as $k \to \infty$ for some $f \in Y$, then some subsequence $x_{n_{k(i)}} \to x$ (resp., there is an $x \in D$) and $Tx = f$.

The pseudo *A*-properness of $T = A–N$ has been established under various assumptions on A and N in [Mi-5-8] and more details with some new examples will be given in Section 2.2. Throughout the paper we shall assume that $(X, \| \cdot \|_0)$ is a Banach space continuously and densely embedded in a Hilbert space.

Without the monotonicity assumption on N, we have

PROPOSITION 2.1 Let X be a reflexive Banach space compactly embedded in a Hilbert space H, $A : D(A) \subset H \to H$ be a closed and densely defined linear map and $N: H \to H$ be a nonlinear continuous map. Then $A + N : D(A) \cap X \subset X \to H$ is pseudo *A*-proper w.r.t.$\Gamma = \{H_n, H_n, P_n\}$ for (X, H) with $P_n Ax = Ax$ on H_n.

Proof. Let $\{x_{n_k} \in X_{n_k}\}$ be bounded in X and $y_k \equiv P_{n_k}(A + N)\, x_{n_k} \to f$ in H. Then we may assume that $x_{n_k} \rightharpoonup x$ (weakly) in X, $x_{n_k} \to x$ in H, and $Nx_{n_k} \to Nx$.

Hence, $Ax_{n_k} = y_k - P_{n_k} Nx_{n_k} \to f - Nx$. Since A is a closed map, $x \in D(A)$ and $Ax + Nx = f$. ∎

Let us first study Eq.(1.1) when there is no resonance at infinity. We have

THEOREM 2.1 Let $A : D(A) \subset H \to H$ be a linear selfadjoint map and $\Gamma=\{ H_n, H_n, P_n \}$ be a projection scheme for (X,H) with $P_n Ax = Ax$ on H_n. Let $N : X \to H$ be a nonlinear selfadjoint map such that for some positive constants a,b,c and r: and $N_\infty : H \to H$

(2.1) *If* $(A - N_\infty)\ x = y$ for $y \in$ H, *then* $x \in X$ and $\| x \|_0 \leq c \| y \|$.

(2.2) $\| Nx - N_\infty\ x \| \quad \leq a \| x \| + b.$ for $\| x \|_0 \geq r$.

(2.3) $0 < a < \min \{ |\mu| \mid \mu \in \sigma\ (A - N_\infty) \}$.

(2.4) Suppose that either one of the following conditions holds:

$$A + N : D(A) \cap X \subset X \to H \text{ is pseudo } A\text{-proper w.r.t.}\Gamma,$$

(2.5) X is a reflexive space compactly embedded in H and N is continuous in H. Then Eq. (1.1) is solvable for each $f \in H$.

Proof. In view of Proposition 2.1, $A + N : D(A) \cap X \subset X \to H$ is pseudo A–proper w.r.t.Γ if (2.5). Therefore, it remains to prove the theorem assuming (2.4).

Now, since $A - N_\infty$ is self–adjoint in H, we have that min $\{|\mu| \mid \mu \in \sigma\ (A - N_\infty)\}$ $\| (A - N_\infty)^{-1} \|^{-1}$. Let $f \in H$ be fixed and $H(t,x)=(A - N_\infty)x - t(N - N_\infty)$ on $[0,1]\ x\ D(A) \cap\ X$. Then there are $\gamma > 0$, $R \geq r$ and $n \geq n_0$ such that

$$\| P_n H(t,x) - t\ P_n\ f \| \quad \geq \gamma \quad \text{for} \quad t \in [0,1], x \in \partial B_X(0,R), n \geq n_0. \tag{2.6}$$

If not, then there would exist $t \in [0,1], t_n \to t_0$, and $x_{n_k} \in H_{n_k}$ such that $\| x_{n_k} \|_0 \to \infty$ and $P_{n_k} H\ (t_k, x_{n_k}) - t_k P_{n_k} f \to 0$ as $k \to \infty$. Then $z_k \equiv y_{n_k} - t_k P_{n_k}(N - N_\infty)\ (A - N_\infty)^{-1} y_{n_k} - t_k\ P_{n_k}\ f \to 0$, where $y_{n_k} = (A - N_\infty) x_{n_k}$. By (2.1) and (2.2) $\| y_{n_k} \| \geq 1/c \| x_{n_k} \|_0 \to \infty$ and

$$\| y_{n_k} \| \leq \delta \| (N - N_\infty)\ (A - N_\infty)^{-1} y_{n_k} \| + \delta \| f \| + \| z_k \| \tag{2.7}$$

$$\leq \delta\ a \| (A - N_\infty)^{-1} \| \quad \| y_{n_k} \| + \delta\ (b + \| f \|) + \| z_k \|.$$

Dividing (2.7) by $\| y_{n_k} \|$ and taking the limit we get $1 \leq \delta a \| \ (A - N_\infty)^{-1} \|$ in contradiction to (2.3). Hence, (2.6) holds and the Brouwer degree $\deg(P_n(A-N), B_X(0,R) \cap X_n, \ P_n f) = \deg(P_n(A-N), B_X(0,R) \cap X_n, \ 0) \neq 0$ for each $n \geq n_0$. Hence, there exists an $x_n \epsilon \ B_X(0,R) \cap X_n$ such that $P_n \ (A-N)x_n = P_n f$ for $n \geq n_0$, and by the pseudo A-properness of $A + N$, there is an $x \ \epsilon \ D(A)$ such that $Ax + Nx = f$. ■

Remark 2.2 If there are real numbers $a < \beta$ such that $\sigma(A) \cap (a, \beta)$ consists of at most a finite number of eigenvalues, and if $\lambda_k < \lambda_{k+1}$ are some consecutive eigenvalues in (a, β) and $\lambda = (\lambda_k + \lambda_{k+1})/2$, then (2.3) holds with $N_\infty = \lambda I$ if $a = \gamma = (\lambda_{k+1} - \lambda_k)/2$. Indeed, the spectral gap for $A - \lambda I$ induced by the gap $(\lambda_k, \lambda_{k+1})$ is $(-\gamma, \gamma)$ and therefore $(A - \lambda I)^{-1} : H \to H$ is a bounded selfadjoint map whose spectrum lies in $(-1/\gamma, 1/\gamma)$. Hence, $\| (A - \lambda I)^{-1} \| = 1/\gamma$.

Regarding condition (2.1), we have the following result useful in applications.

LEMMA 2.1 Let $A : D(A) \subset H \to H$ be selfadjoint with the spectum $\sigma(A)$ consisting only of eigenvalues $\{\lambda_i | i \epsilon \mathrm{I}\}$ having no accumulation points and each $\{\lambda_i \neq 0\}$ have a finite multiplicity j_i. Suppose that the corresponding eigenvectors $\{e_{ij} | i \epsilon \mathrm{I}, \ 1 \leq j \leq j_i\}$ form a complete basis for X and H. Suppose that there is a constant $c_0 > 0$ such that if $Ax = y$ for some $y \epsilon \ker A$, then $x \epsilon X$ and $\| x \|_0 \leq c_0 \| y \|$. Then condition (2.1) holds.

Proof. For $x \epsilon D(A)$ we have

$$x = \sum_i \sum_{j=1}^{j_i} x_{ij} \, e_{ij} \quad \text{and} \quad Ax = \sum_i \lambda_i \sum_{j=1}^{j_i} x_{ij} \, e_{ij},$$

where $x_{ij} = (x, \ e_{ij})$. If $\lambda \notin \sigma(A)$ and $(A - \lambda \mathrm{I})x = y$ for $y \epsilon H$ then

$$(A - \lambda \mathrm{I})x = \sum_i (\lambda_i - \lambda) \sum_{j=1}^{j_i} x_{ij} e_{ij} = \sum_i \sum_{j=1}^{j_i} y_{ij} e_{ij}$$

and therefore

$$x = (A - \lambda \mathrm{I})^{-1} y = \sum_i \frac{1}{\lambda_i - \lambda} \sum_{j=1}^{j_i} y_{ij} \, e_{ij} \ .$$

Since $\ker(A - \lambda \mathrm{I}) = \{0\}$, there is a constant $a > 0$ such that

$|\lambda_i - \lambda| \geq a$ for all $i\epsilon I$. If not, then a subsequent $\lambda_{i_k} \to \lambda$ in contradiction to the fact that $\{\lambda_i\}$ has no accumulation points. Since $x_{ij} = y_{ij}/(\lambda_i - \lambda)$, we have that

$$|x_{ij}| \leq a^{-1}|y_{ij}| \quad \text{and} \quad \| x \| \leq c_1 \| y \|$$

for some $c_1 > 0$ independent of y. Moreover, $x\epsilon X$ and $\| x \|_0 \leq c_0 \| y + \lambda x \|$ since $Ax = y + \lambda x$. Hence, $\| x \|_0 \leq c(\lambda) \| y \|$ with $c(\lambda) = c_0 + \lambda c_1$. ■

If A is not selfadjoint, analyzing the proof of Theorem 2.1 we see that the following more general version of it holds.

THEOREM 2.2 Let $C : X \to H$ be a linear map such that A–C: $D(A) \cap X \to H$ is a bijection and for some positive constants a, b, c, $\delta = \max \| P_n \|$ and r with $ac\delta < 1$:

(2.8) $\| (A - C)^{-1} y \|_0 \leq c \| y \|$ for each $y\epsilon H$,

(2.9) $\| Nx - Cx \| \leq a \| x \|_0 + b$ for each $\| x \|_0 \geq r$.

Then the conclusions of Theorem 2.1 are valid.

Finally, we shall prove an abstract result suitable for studying hyperbolic equations with nonlinear dissipation. It is motivated by Prodi's work [Pr] on nonlinear telegraph equations.

THEOREM 2.3 Let $A : D(A) \subset H \to H$ be a closed densely defined linear map and $M : X \to H$ and $N : H \to H$ be nonlinear maps such that

(2.10) There are a map $C : X \to H$ and $\lambda_1 > 0$ such that $(Ax, \ Cx) \geq 0$ and

$$(Ax, x) \geq \lambda_1 \| x \|^2 - \| Cx \|^2 \quad \text{for } x\epsilon D(A) \cap X.$$

(2.11) There are a map $S : X \to H$ with $(Sx, \ x) = 0$ for $x\epsilon X$ and constants $a > 0$, $r > 0$ and $m > 0$ such that
$\| Mx - Sx \| \leq a \| Cx \|$, $(Mx, Cx) \geq m \| Cx \|^2$ for $\| x \|_0 \geq r$.

(2.12) There are a map $R : H \to H$ with $(Rx, x) = 0$ for $x\epsilon X$, a vector $y\epsilon H$ and constants $b \geq 0$, $c \geq 0$ and $d > 0$ such that

$\| Nx - Rx \| \leq b \| x \| + c$ for $x \epsilon H$,

(2.13) $(Nx, x) \leq d \| x \|^2 + (y, x)$ for $x \epsilon H$,

(2.14) $m^2(\lambda_1 - d) - mab - b^2 > 0$.

Then if $H(t,x) = Ax + Mx - tNx$ is an A–proper homotopy $w.r.t.\Gamma = \{H_n, P_n\}$ for H with $P_n Ax = Ax$ on H_n and $\deg(P_n H_0, B(0,R) \cap H_n, 0) \neq 0$ for $n \geq n_0$ and all large R, the equation $Ax + Mx - Nx = f$ is solvable in H for each $f \in H$.

Proof. Let $f \epsilon H$ be fixed. We shall show first that all solutions $x \epsilon X$ of

$$Ax + Mx - Nx = f \tag{2.15}$$

are bounded in H. Taking the scalar product of (2.15) with Cx and then with x, we get respectively

$$(Mx, Cx) \leq (Nx, Cx) + (f, Cx), \tag{2.16}$$

$$\lambda_1 \| x \|^2 - \| Cx \|^2 + (Mx, x) - (Nx, x) - (f, x) \leq 0. \tag{2.17}$$

Multiplying (2.16) by $\varrho > 0$ and adding it to (2.17) we get,

$$\varrho(Mx, Cx) - \| Cx \|^2 - \varrho(Nx, Cx) + (Mx, x) + \lambda_1 \| x \|^2 - (Nx, x) - \varrho(f, x) \leq 0 \tag{2.18}$$

Using the properties of the maps involved, we get

$$(\varrho m - 1) \| Cx \|^2 - (\varrho b + a) \| Cx \| \; \| x \| + (\lambda_1 - d) \| x \|^2$$

$$- \varrho c \| Cx \| - (\| y \| - \varrho \| f \|) \| x \| \leq 0. \tag{2.19}$$

The discriminant for the quadratic part in $\| x \|$ and $\| Cx \|$ in (2.19)

$$D(\varrho) = 4(\varrho m - 1)(\lambda_1 - d) - (\varrho b + a)^2$$

is easily seen to have a positive maximum for some $\varrho_0 > 0$ since $\lambda_1 > d$ by (2.12) when $b = 0$.

To obtain the boundedness of all solutions of (2.15) from (2.19), consider the function of two real variables (ξ, η):

$$\psi(\xi,\eta) = A\xi^2 + 2B\xi\eta + C\eta^2 + 2D\xi + 2E\eta .$$

Suppose that the corresponding quadratic form $A\xi^2 + 2B\xi\eta + C\eta^2$ is positive definite. Since $\psi(\xi,\eta) = 0$ is an ellipse passing through the origin, its interior points satisfy $\psi(\xi,\eta) \le 0$. Since the distance of (ξ,η) on the ellipse to the origin is less than or equal to the major axis of the ellipse, using elementary calculus we see that

$$\xi^2 + \eta^2 \le \frac{16F^2}{(AC - B^2)^2}(D^2 + E^2),$$

where $F = \max\{A, |B|, C\}$. For ϱ such that $D(\varrho) > 0$ this estimate applied to (2.19) gives

$$\| x \|^2 + \| Cx \|^2 \le \frac{64F^2(\varrho)}{[4(\varrho m - 1)(\lambda_1 - d) - (\varrho b + a)^2]^2}\left((\| y \| + \varrho \| f \|)^2 + c^2\varrho^2\right),$$

where

$$F(\varrho) = \max\{\varrho m - 1, (\varrho b + a)/2, \lambda_1 - d\} .$$

Hence, all solutions $x \epsilon X$ of (2.15) lie in a ball $B(0,R) \subset H$ for some $R \ge r$ large enough.

Next, we need to show that all solutions $x \epsilon X$ of

$$Ax + Mx - tNx = tf, \quad 0 \le t \le 1, \tag{2.20}$$

are bounded in H. This has been shown above for t=1. Since $m^2\lambda_1 > 0$, by (2.14) we get

$$m^2(\lambda_1 - td) - mtab - t^2b^2 > 0, \quad 0 \le t \le 1. \tag{2.21}$$

Then, as for $t=1$, we get for any solution x of (2.20) for fixed t,

$$\| x \|^2 + \| Cx \|^2 \leq \frac{64F^2(\varrho, t)}{[4(\varrho m - 1)(\lambda_1 - td) - (tb\varrho + a)^2]^2} \, t^2 \left((\| y \| + \varrho \| f \|)^2 + c^2\varrho^2 \right),$$

where $F(\varrho, t) = \max\{\varrho m - 1, (tb\varrho + a)/2, \lambda_1 - td\}$. As before, for each t fixed, the function

$$D(\varrho; t) = 4(\varrho m - 1)(\lambda_1 - td) - (t\varrho b + a)^2$$

has a positive maximum at some $\varrho_0(t) > 0$ by (2.21). Moreover, an easy calculation shows, by (2.21), that there is a constant $\delta > 0$ such that

$$D(\varrho; t) \geq \delta \text{ for } all \ 0 \leq t \leq 1, \tag{2.22}$$

and the set of all $\varrho's$ for which (2.22) holds is bounded with each $\varrho > 0$. Actually, it can be shown that if $\delta > 0$ is sufficiently small then as ϱ we can take the smaller root of $D(\varrho; t) = \delta$.

Now, from the form of $F(\varrho; t)$ we see $\| x \| < R$ for all solutions of (2.20) with $R \geq r$ sufficiently large. Hence, $H(t, x) = Ax + Mx - tNx \neq tf$ for all $\| x \| = R$ and $0 \leq t \leq 1$. Since $H(t,x)$ is an A-proper homotopy in H w.r.t. some $\Gamma = \{H_n, P_n\}$ for H, arguing by contradiction it is easy to show that there is an $n_0 \geq 1$ such that

$$P_nH(t, x) \neq tP_nf \text{ for all } x \in \partial B(0, R) \cap H_n, \ t \in [0, 1], n \geq n_0.$$

Hence, by the homotopy theorem for Brouwer's degree, there is an $x_n \in B(0, R) \cap H_n$ such that $P_n(A + M - N)x_n = P_nf, n \geq n_0$, since $\deg(P_nH_1, B(0, R) \cap H_n, P_nf) = \deg(P_n(A + M), B(0, R) \cap H_n, 0) \neq 0$. By the A-properness of $A+M-N$, a subsequence of $x_{n_k} \to x$ and $Ax + Mx - Nx = f$. ■

COROLLARY 2.1 Let X be compactly embedded in H and conditions (2.10) - (2.14) hold. Suppose that $M : X \to H$ is weakly continuous (i.e. $Mx_n \rightharpoonup Mx$ in H if $x_n \rightharpoonup x$ in X), $N : H \to H$ is bounded and demicontinuous (i.e., $Nx_n \rightharpoonup Nx$ in H if $x_n \to x$ in H) and there are constants $c_0, c_1 > 0$ such that

$$\| P_n(Ax + Mx) \| \geq c_0 \| x \|_0 - c_1 \| x \| \text{ for } x \in X_n, n \geq n_0. \tag{2.23}$$

Then, if deg $(P_n(A+M), B(0,R)\cap H_n, 0) \neq 0$ for all $n \geq n_0(R)$ and all R large with $P_nA(H_n) \subset H_n$, the equation $Ax+Mx-Nx = f$ is solvable for each $f\epsilon H$.

Proof. In view of the proof of Theorem 2.3, it remains only to prove that $H(t, x) = Ax+Mx-tNx$ is an A-proper homotopy w.r.t. $\Gamma = \{H_n, P_n\}$ on $[0,1]\ x\ \bar{B}(0,R)$. Since $H_n \subset D(A)\cap X$ and all norms are equivalent in H_n, it follows that $P_nM : H_n \subset H \to H_n$ is continuous. Similarly, $P_nN : H_n \to H_n$ is continuous, and therefore $P_nH(t,\cdot) : H_n \to H_n$ is continuous for each $t\epsilon[0,1]$.

Next, suppose that $\{x_{n_k}\epsilon H_{n_k}\}$ is bounded and $y_{n_k} \equiv P_{n_k}H(t_k, x_{n_k}) \to f$ as $k \to \infty$. By (2.23),

$$c_0 \| x_{n_k} \|_0 \leq \| y_{n_k} \| + \| Nx_{n_k} \| + c_1 \| x_{n_k} \| \leq C \text{ for all } k$$

and some constant C. Hence, we may assume that $x_{n_k} \to x$ in H, with $x\epsilon X$, and since $P_nAx = Ax$ *on* H_n and A is a weakly closed map, $P_{n_k}(Ax_{n_k} + Mx_{n_k} - tNx_{n_k}) \rightharpoonup Ax + Mx - Nx = f$. Thus, $H(t,x)$ is an *A*-proper homotopy w.r.t.Γ. ■

Remark 2.3 It is well known that deg $(P_n(A+M), B(0,R)\cap H_n, 0) \neq 0$ for $n \geq n_0$ if, e.g. $T_n = P_n(A+M) : H_n \to H_n$ is injective, or M is an odd map (i.e., $M(-x) = -Mx$ on $X\backslash B(0, R_0))$ or $T_nx, Bx) > -\| T_nx \| \| Bx \|$ for $\| x \| = R$ and some continuous map $B : H_n \to H_n$ with deg $(B, B(0,R)\cap H_n, 0) \neq 0$ (cf. [Mi-4]). In particular, if M is strictly C-monotone, i.e. $(Mx-My, Cx-Cy) > 0$ for all $x, y\epsilon X, x \neq y$, then such is $A+M$ since $(Ax, Cx) = 0$ on $X\cap D(A)$. Hence, if $P_nC(H_n) \subset H_n$, then $P_n(A+M)$ is strictly C-monotone on H_n and is therefore injective.

2.2 Semilinear Equations with $\{B_1, B_2\}$-quasilinear Perturbations.

We continue our study of Eq. (1.1) with the so-called asymptotically $\{B_1, B_2\}$ - quasilinear nonlinearities N. Such maps have been introduced by Perov [Per] and Krasnoselskii–Zabreiko [Kr–Za] in their study of the existence of fixed points of compact maps. Let $B_1, B_2 : H \to H$ be selfadjoint maps with $B_1 \leq B_2$, i.e. $(B_1x, x) \leq (B_2x, x)$ for $x \in H$.

Definition 2.2 a) A nonlinear map $K : H \to H$ is $\{B_1, B_2\}$-quasilinear on a set $S \subset H$ if for each $x \in S$ there exists a linear selfadjoint map $B : H \to H$ such that $B_1 \leq B \leq B_2$ and $Bx = Kx$;

b) A map $N : H \to H$ is said to be asymptotically $\{B_1, B_2\}$-quasilinear if there is a $\{B_1, B_2\}$-quasilinear outside some ball map K such that

$$|N - K| = \lim_{\|x\| \to \infty} \sup \frac{\| Nx - Kx \|}{\| x \|} < \infty .$$

We do not require that $\{B_1, B_2\}$ is a regular pair as in [Kr–Za, Per]. This class of maps, is rather large. For example, let $K : H \to H$ have a selfadjoint weak Gateaux derivative N', i.e. $N'(x)$ is a selfadjoint map on H for each x and

$$\lim_{t \to 0} t^{-1}(N(x + th) - N(x), y) = (N'(x)h, y)$$

for all $x, y, h \in H$. Assume that $B_1 \leq N'(x) \leq B_2$ for each x and some selfadjoint maps B_1 and B_2. Then N is asymptotically $\{B_1, B_2\}$-quasilinear with $|N - K| = 0$. Indeed, for every $x, y, t \in H$, the mean value theorem implies the existence of a number $t \in (0, 1)$ such that

$$(Nx - Ny, z) = (N'(y + z(x - y))(x - y), z).$$

When $y = 0$, this gives

$$(Nx - N'(tx)x, z) = (N(0), z)$$

and therefore, if we set $Kx = N'(tx)x$, then

$$\| Nx - Kx \| = \sup_{\|z\| \leq 1} |(N(0), z)| \leq \| N(0) \| .$$

Hence, $|N-K| = 0$ and N is asymptotically $\{B_1, B_2\}$-quasilinear. In the nondifferentiable case, if $Nx = B(x)x + Mx$ for some nonlinear map M with the quasinorm $|M| < \infty$ and selfadjoint maps $B(x) : H \to H$ with $B_1 \leq B(x) \leq B_2$ for each x in H, then N is asymptotically $\{B_1, B_2\}$-quasilinear.

Next , we shall prove some preliminary results.

LEMMA 2.3 Let $A : D(A) \subset H \to H$ and $B^{\pm} \epsilon\ L(H)$ be selfadjoint with $B^- \leq B^+$ and $H^{\pm}$ be subspaces with $H = H^- \oplus H^+$ and such that for some $\gamma_1 > 0$ and $\gamma_2 > 0$

$$((A - B^-)x, x) \leq -\gamma_1 \| x \|^2 \quad \text{for all} \quad x \epsilon\ H^- \cap D(A) \tag{2.24}$$

$$((A - B^+)x, x) \geq \gamma_2 \| x \|^2 \quad \text{for all} \quad x \epsilon H^+ \cap D(A) \tag{2.25}$$

Then there are $\epsilon > 0$ and $c > 0$ such that for any selfadjoint maps B_1, B_2, $C \in L(H)$ with $B_1 \leq B^-$ and $B^+ \leq B_2$ and

$$B_1 - \epsilon I \leq C \leq B_2 + \epsilon I \tag{2.26}$$

we have that

$$\| Ax - Cx \| \geq c \| x \| \quad \text{for} \quad x \in D(A). \tag{2.27}$$

Proof. If (2.27) does not hold, then there would exist selfadjoint maps $B_{1,n}$, $B_{2,n}$ and $C_n \epsilon\ L(H)$ and $x_n \epsilon\ D(A)$ with $\| x_n \| = 1$ such that $B_{1,n} \leq B^-$, $B^+ \leq B_{2,n}$

$$B_{1n} - I/n \leq C_n \leq B_{2n} + I/n \tag{2.28}$$

and

$$\| Ax - C_n x_n \| \leq \| x_n \| / n \,.$$

Set $y_n = Ax_n - C_n x_n$. Then, we have $x_n = x_{1n} + x_{2n} \in H^- \oplus H^+$ and

$$(Ax_{2n} - C_n x_{2n}, x_{2n}) - (Ax_{1n} - C_n x_{1n}) = (y_n, x_{2n} - x_{1n}) \ .$$

Since $B_{1,n} \leq B^-$ and $B^+ \leq B_{2,n}$, (2.28) and (2.26) through (2.27) imply that

$$(Ax_{1n} - C_n x_{1n}, x_{1n}) \leq (Ax_{1n} - B^- x_{1n}, x_{1n}) + \| x_{1n} \|^2 / n \leq (1/n - \gamma_1) \| x_{1n} \|^2 \,.$$

Subtracting the first equation from the second one, we get

$$\gamma_1 \| x_{1n} \|^2 + \gamma_2 \| x_{2n} \|^2 - (\| x_{1n} \|^2 + \| x_{2n} \|^2)/n$$
$$\leq (Ax_{2n} - C_n x_{2n}, x_{2n}) - (Ax_{1n} - C_n x_{1n}, x_{1n})$$
$$= (y_n, x_{2n} - x_{1n}) \leq \| x_{2n} - x_{1n} \| /n.$$

Hence, if $\gamma = \min\{\gamma_1, \gamma_2\}$, by the parallelogram law we get

$$(\gamma - 1/n)(\|x_{2n}\|^2 + \| x_{1n} \|^2) \leq 2/n(\| x_{1n} \|^2 + \| x_{2n} \|^2)$$

and therefore $\gamma \leq 1/n \to 0$ as $n \to \infty$, in contradiction to $\gamma > 0$. Hence, (2.27) holds. ∎

Now we are ready to prove our basic solvability result for Eq. (1.1) involving asymptotically $\{B_1, B_2\}$-quasilinear nonlinearities N without resonance at infinity. It is based on the following continuation theorem.

THEOREM 2.4 [Mi–3] Let V be dense subspace of a Hilbert space H, $D \subset H$ be open and bounded subset and a homotopy $H : [0,1]x(\bar{D} \cap V) \to H$ be such that
(i) H is an A–proper homotopy w.r.t.$\Gamma = \{H_n, P_n\}$ on $[0, \epsilon]$ x $(\partial D \cap V)$ for each ϵ in $(0,1)$, and H_1 is pseudo A–proper w.r.t.Γ

(ii) $H(t,x)$ is continuous at 1 uniformly for $x \in \bar{D} \cap V$.

(iii) $H(t,x) \neq f$ and $H(0,x) \neq tf$ for $t \in [0,1]$, $x \in \partial D \cap V$.

(iv) $deg\ ((P_n H_o, D \cap H_n, 0) \neq 0$ for all large n.

Then the equation $H(1, x) = f$ is solvable in $\bar{D} \cap V$.

Now, we have

THEOREM 2.5 Let $A : D(A) \subset H \to H$ satisfy (2.24) – (2.25) and $N : H \to H$ be bounded asymptotically $\{B_1, B_2\}$–quasilinear with $|N–K|$ sufficiently small. Suppose that a selfadjoint map $C_0 \in L(H)$ satisfies (2.26) with sufficiently small ϵ and that $H(t, \cdot) = A - (1-t)C_0 - tN$ is A–proper w.r.t.$\Gamma = \{H_n, P_n\}$ with $P_n Ax = Ax$ on H_n for each $t \in [0,1)$ and H_1 is pseudo A–proper w.r.t.Γ. Then Eq. (1.1) is solvable for each $f \in H$.

Proof. Since $N_f x = Nx - f$, $f \in H$, has the same properties as N, it suffices to solve the equation $Ax - Nx = 0$. Let $\epsilon_0 > 0$ be such that $|N - K| + \epsilon_0 < c$, where c is from Lemma (2.3). Then there is an $r > 0$ such that

$$\| Nx - Kx \| \leq (|N - K| + \epsilon_0) \| x \| \quad \text{for each} \quad \| x \| \geq r\,.$$

Moreover,

$$H(t,x) = Ax - (1-t)C_0x - tNx \neq 0 \quad \text{for} \quad x \in \partial B_r \cap D(A), t \in [0,1]. \tag{2.29}$$

If not, then $H(t,x) = 0$ for some $\| x \| = r$ and $t \in [0,1]$. Hence, subtracting tKx from both sides, we get

$$\| Ax - tKx - (1-t)C_0x \| = t \| Nx - Kx \| < c \| x \|.$$

Since K is $\{B_1, B_2\}$-quasilinear, there is a selfadjoint map $C_* \in L(H)$ such that $Kx = C_*x, B_1 \leq C_* \leq B_2$ and therefore

$$\| Ax - tC_*x - (1-t)C_0x \| < c \| x \|\,. \tag{2.30}$$

But, $C = tC_* + (1-t)C_0$ is selfadjoint satisfies (2.26) in Lemma 2.3 and therefore (2.27) holds. This contradicts (2.30) and so (2.29) is valid.

Next, since C_0 and N are bounded maps, $H(t,x)$ is an A-proper homotopy on $x \in \bar{B}_r \cap D(A)$ w.r.t.Γ for each ϵ in $(0,1)$ and is continuous at **1** uniformly for $x \in \bar{B}_r \cap D(A)$. Hence, the solvability of $Ax - Nx = 0$ follows from Theorem 2.4. ■

Let us now discuss some conditions on $B^{\pm}$ which imply (2.24) – (2.25). Assume, as in Amann [Am–2],

(2.31)

a) $A : D(A) \subset H \to H$ is selfadjoint

b) $B^{\pm} = \sum_{i=1}^{m} \lambda_i^{\pm} P_i^{\pm}$ commute with A, where $P_i^{\pm} : H \to Ker(B^{\pm} - \lambda_i)$ are orthogonal projections, $\lambda_i^{\pm} \leq \cdots \leq \lambda_m^{\pm}$ and $\lambda_i^{\pm}$ are pairwise distinct.

c) $\bigcup_{i=1}^{m} [\lambda_i^{-}, \lambda_i^{+}] \subset \varrho(A)$ – the resolvent set of A.

Being selfadjoint, A possesses a spectral resolution

$$A = \int_{-\infty}^{\infty} \lambda d\mathrm{E}_\lambda,$$

where $\{ \mathrm{E}_\lambda \mid \lambda \epsilon R \}$ is a right continuous spectral family. Since $B^{\pm}$ commute with A, it is known that $P_i^{\pm}$ commute with the resolution of the identity $\{E_\lambda | \lambda \in R\}$. Hence, the selfadjoint maps $A - B^{\pm}$ have the spectral resolution

$$A - B^{\pm} = \sum_{i=1}^{m} \int_{-\infty}^{\infty} (\lambda - \lambda_i^{\pm}) dE_\lambda \; P_i^{\pm}. \tag{2.32}$$

Define the orthogonal projections $P^{\pm}$ by

$$P^{-} = \sum_{i=1}^{m} E(-\infty, \lambda_i^{-})P_i^{-} \quad \text{and} \quad P^{+} = \sum_{i=1}^{m} E(\lambda_i^{+}, \infty)P_i^{+},$$

where

$$E(a, \beta) = \int_a^\beta dE_\lambda$$

for all $a, \beta \in \varrho(A) \cup \{\pm \infty\}$ with $a < \beta$. Define $H^{\pm} = P^{\pm}(H)$ and note that by (2.31) – (c),

$$P^{+} = \sum_{i=1}^{m} E(\lambda_i^{-}, \infty)P_i^{+}$$

and

$$\gamma = dist(\bigcup_{i=1}^{m} [\lambda_i^{-}, \lambda_i^{+}], \sigma(A)) > 0.$$

Moreover, by (2.32), we have that

$$\big((A - B^{-})x, x\big) \leq -\gamma \| x \|^2 \quad \text{for} \quad x \epsilon \; D(A) \cap H^{-},$$

$$\big((A - B^{-})x, x\big) \geq -\gamma \| x \|^2 \quad \text{for} \quad x \epsilon \; D(A) \cap H^{+}.$$

Hence, we have

LEMMA 2.4 If (2.31) holds and $P_i^- = P_i^+$ for $1 \le i \le m$, then there are orthogonal subspaces $H^{\pm}$ such that $H = H^- \oplus H^+$ and conditions (2.24) – (2.25) hold with $\gamma_1 = \gamma_2 = \gamma > 0$.

Proof. It remains only to show that H^+ and H^- are orthogonal. Since $P_i^- = P_i^+$ for $i = 1, \ldots, m$, and $E(-\infty, \lambda_i^-) = I - E(\lambda_i^-, \infty)$, we get that $P^+ = I - P^-$ and therefore, $H^+ = (H^-)^{\perp}$. ■

When $B^{\pm}$ are not of the form (2.31) (–b), we need to assume more on the linear part A.

(2.33) Suppose A is selfadjoint possessing a countable spectrum $\sigma(A)$ consisting of eigenvalues and whose eigenvectors form a complete orthonormal system in H.

(2.34) There are selfadjoint maps $C_1, C_2 \in L(H)$ and two consecutive finite multiplicity eigenvalues $\lambda_k < \lambda_{k+1}$ of A such that

$$\lambda_\kappa \| x \|^2 < (C_1 x, x) \le (C_2 x, x) < \lambda_{\kappa+1} \| x \|^2 \quad \text{for } x \in H \backslash \{0\}.$$

Let H^- (resp. H^+) be the subspaces of H spanned by the eigenvectors of A corresponding to the eigenvalues $\lambda_i \le \lambda_\kappa$ (resp., $\lambda_i \ge \lambda_{\kappa+1}$).

LEMMA 2.5 Let (2.33) – (2.34) hold. Then there are $\gamma_1 > 0$ and $\gamma_2 > 0$ such that for any selfadjoint maps $B^{\pm} \in L(H)$ satisfying $C_1 \le B^-$ and $B_+ \le C_2$ on H, we have that (2.24) – (2.25) hold.

Proof. It is enough to show that (2.24) holds since the same arguments give also (2.25). Since $\lambda_\kappa < C_1 \le B^-$, it is enough to show that there is a $\gamma_1 > 0$ such that

$$(Ax - C_1 x, x) \le -\gamma_1 \| x \|^2 \quad \text{for all } x \in D(A) \cap H^-.$$

If such a $\gamma_1 > 0$ did not exist, then there would exist $\{x_n\} \subset D(A) \cap H^-$ with $\| x_n \| = 1$ and such that

$$-1/n \le (Ax_n - C_1 x_n, x_n) \le (Ax_n - \lambda_k x_n, x_n), \quad n = 1, 2, \ldots.$$

Decompose $H^- = \tilde{H} \oplus \bar{H}$, where $\tilde{H}$ is spanned by the eigenvaectors $\{e_i\}$ corresponding to $\lambda_i \leq \lambda_{k+1}$ and $\bar{H}$ is the finite dimensional space spanned by the eigenvectors corresponding to λ_k. Then $x_n = \tilde{x}_n + \bar{x}_n \in \tilde{H} \oplus \bar{H}$ and

$$- 1/n \leq (Ax_n - \lambda_k x_n, x_n) = (A\tilde{x}_n - \lambda_k \tilde{x}_n, \tilde{x}_n)$$

$$= \sum_{i \leq k-1} \lambda_i (\tilde{x}_n, e_i)^2 - \lambda_k \| \tilde{x}_n \|^2 \leq (\lambda_{k-1} - \lambda_k) \| \tilde{x}_n \|^2$$

Hence, $\tilde{x}_n \to 0$ as $n \to \infty$ and $x_n \to \bar{x} \in \bar{H}$ with $\|\bar{x}\| = 1$ since $\| x_n \|^2 = 1 - \| \tilde{x}_n \|$ and $\bar{H}$ is finite dimensional. Thus,

$$- 1/n \leq (\lambda_k x_n - C_1 x_n, x_n) + (Ax_n - \lambda_k x_n, x_n)$$

$$\leq (\lambda_k x_n - C_1 x_n, x_n) + (\lambda_{k-1} - \lambda_k) \| \tilde{x}_n \|^2$$

and passing to the limit as $n \to \infty$, we get

$$0 \leq ((\lambda_k - B_1)\, \bar{x}, \bar{x}), \quad \bar{x} \neq 0$$

in contradiction to (2.34). Hence, (2.24) holds. ■

Remark 2.4 If λ_i (resp., λ_{i+1}) is of infinite multiplicty, then Lemma 2.4 is still valid if we assume in (2.34) :

$$(\lambda_i + \epsilon) \| x \|^2 \leq (C_1 x, x) \ \ (resp., (C_2 x, x) \leq (\lambda_{i+1} - \epsilon) \| x \|^2) \ \text{ for } \ 0 \neq x \in H.$$

This is easy to check by analyzing the proof of this lemma.

Next, let us look at the case when $H^- \oplus H^+ \neq H$, which is also useful in applications.

Recall that a closed subspace $X \subset H$ is said to reduce A if A commutes with the orthogonal projection P of H onto X, i.e., if $PA \subset AP$. As before, let $P^{\pm}$ be the orthogonal projections of H onto $H^{\pm}$. Regarding A and H we require that the following conditions hold for a scheme $\Gamma = \{H_n, P_n\}$ (cf. [Am–2]):

(2.35) (i) $H_1 \subset H_2 \subset \dots$ with each H_n closed in H, $\dim H_n = \infty$ and $\bigcup H_n$ is dense in H;

(ii) each H_n reduces A;

(iii) the orthogonal projections $P_n : H \to H_n$ commute with; $P^\pm$

(iv) $H_n = H_n^- \oplus H_n^+$, where $H_n^\pm = H^\pm \cap H_n$;

(v) $Q^\pm(D(A) \cap H_n) \subset D(A)$ for each n, where $Q^\pm : H^- \oplus H^+ \to H^\pm$ are orthogonal projections.

We have the following extension of Theorem 2.5.

THEOREM 2.6 [Mi–8] Let $H^\pm$ be closed subspaces of H such that $H^- \cap H^+ = \{0\}, A : D(A) \subset H \to H$ satisfy conditions (2.24) – (2.25) and (2.35) hold. Suppose that $N : H \to H$ is bounded asymptotically $\{B_1, B_2\}$-quasilinear with $|N–K|$ sufficiently small and $A - N : D(A) \subset H \to H$ is pseudo A–proper *w.r.t.* $\Gamma = \{H_n, P_n\}$. Suppose that a selfadjoint map $C_0 \in L(H)$ satisfies (2.26) and for each large n, the map $H_n(t, .) = P_n(A - (1-t)C_0 - tN) : D(A) \cap H_n, \to H_n$ is A–proper $w.r.t.\Gamma_n = \{H_{n,k}, Q_k\}$ for H_n, with $Q_k Ax = Ax$ on $H_{n,k}$, for each $t \in [0, 1)$ and $H_n(1, .)$ be pseudo A–proper $w.r.t.\Gamma_n$. Then Eq. (1.1) is solvable for each $f \in H$.

When $P_i^- \neq P_i^+$ for $1 \leq i \leq m$ and $H \neq H^- \oplus H^+$, regarding conditions (2.24) – (2.25) we have (cf. [Am–2]).

LEMMA 2.6 Let (2.31) hold and there exist a unitary map $U \in L(H)$ such that A commutes with U and

$$P_i^- = UP_i^+U^{-1}, i = 1, \dots, m.$$

Suppose that A has a pure point spectrum in $(\lambda_1^-, \lambda_m^+)$. Then there are closed subspaces $H^\pm$ of H with $H^- \cap H^+ = \{0\}$ such that conditions (2.24) – (2.25) hold with $\gamma_1 = \gamma_2 = \gamma > 0$, and a scheme $\Gamma = \{H_n, P_n\}$ satisfying (2.35).

To discuss some special cases of Theorem 2.5, we shall exhibit several new classes of (pseudo) A–proper maps. Recall that $N : H \to H$ is monotone if $(Nx - Ny, x - y) \geq 0$ for all $x, y \in H$. It is said to be of type (M) if whenever $x_n \rightharpoonup x$ in H and *limsup* $(Nx_n, x_n - x) \leq 0$, then $Nx_n \rightharpoonup Nx$. It is known (cf.

[P–S]) that bounded monotone maps are of type (M). We say that N is demicontinuous if $x_n \to x$ in H implies that $Nx_n \rightharpoonup Nx$.

For such nonlinearities, we have

PROPOSITION 2.2 [Mi–5,6] Let $A : D(A) \subset H \to H$ be a closed linear densely defined map with $R(A) = \ker A^{\perp}$ and $N : H \to H$ be bounded and of type (M). Suppose that either one of the following conditions holds:

(2.36) $A^{-1} : RA) \to R(A)$ is compact,

(2.37) $A = A^*, 0 \epsilon \sigma(A)$ and $\sigma(A) \cap (0, \infty) \neq \phi$ and consists of isolated eigenvalues having finite multiplicities.

Then $\pm A + N$ is pseudo A–proper w.r.t.$\Gamma = \{H_n, P_n\}$ for H with $P_n Ax = Ax$ on H_n.

Proof. Suppose first that (2.36) holds. Let $\{x_{n_k} \in H_{n_k}\}$ be bounded and $y_k = Ax_{n_k} - P_{n_k}Nx_{n_k} \to f$ in H as $k \to \infty$. Then we may assume that $x_{n_k} = x_{0_{n_k}} + x_{1_{nk}} \rightharpoonup x$, $x_{0_{n_k}} = Px_{n_k}$, $x_{1n_k} = (I - P)x_{n_k}$, and

$$PP_{n_k}Nx_{n_k} \to Pf, \quad z_k \equiv Ax_{1n_k} - (I - P)P_{n_k} Nx_{n_k} \to (I - P)f. \tag{2.38}$$

Since N is bounded, so is $\{ Ax_{n_k}\}$ and we may assume that $Ax_{n_k} \rightharpoonup y$. By the weak closedness of the graph of A, it follows that $x \epsilon D(A)$ and $Ax = y$. Moreover, the second equation in (2.38) implies that $x_{1n_k} \to x_1$ and so $x_o \epsilon \ker A$.

Next, by (2.38)

$$\begin{aligned}
&(Nx_{n_k}, x_{n_k} - x) = (P_{n_k}Nx_{n_k}, x_{n_k} - x) - (Nx_{n_k}, (I - P_{n_k})x) \\
&= (PP_{n_k}Nx_{n_k}, x_{n_k} - x) + ((I - P)P_{n_k}Nx_{n_k}, (I - P)(x_{n_k} - x)) \\
&- (Nx_{n_k}, (I - P_{n_k})x) \to 0 \quad \text{as } k \to \infty.
\end{aligned} \tag{2.39}$$

Hence, since N is of type (M), $P_{n_k}Nx_{n_k} \rightharpoonup Nx$ and so $Ax_{n_k} \rightharpoonup f - Nx = Ax$. Thus, $Ax - Nx = f$.

Next, suppose that (2.37) hold. By the above discussion, we need only to show that *limsup* $(Nx_{n_k}, x_{n_k} - x) \leq 0$, where $\{ x_{n_k}\}$ is as above. To that end, let λ_1 be the smallest positive eigenvalue of A and $K : R(A) \to R(A)$ be the right inverse of $-A$, i.e. $K = (-A|D(A) \cap R(A))^{-1}$. Then, by the closed graph theorem, K is a bounded linear map on $R(A)$ and $\mu\epsilon\sigma(K)\backslash\{0\}$, if and only if, $-1/\mu\epsilon\sigma(A)$. Let $\{E_\lambda|\lambda\epsilon R\}$ be the spectral resolution of $-A$, and set

$$P^- = \int_{-\infty}^{-\lambda_{1/2}} dE_\lambda \ , \quad P^+ = \int_{-\lambda_1/2}^{\infty} dE_\lambda \ , \quad H^\pm = P^\pm(R(A)). \tag{2.40}$$

Then $P^\pm$ are orthogonal projections, $R(A) = H^+ \oplus H^-$ (orthogonal direct sum), $KH^\pm \subset H^\pm, KP^+$ is semi-positive definite on $R(A)$ and, by (2.37), KP^- is compact on $R(A)$ and $(Kx, x) \geq -1/\lambda_1 \ \|x\|^2$ for $x\epsilon R(A)$. Since KP^- is compact, we have

$$P^-x_{n_k} = P^-K(-A)x_{n_k} = -KP^-Ax_{n_k} \to -KP^-Ax = P^-x.$$

Since

$$(I-P)P_{n_k}Nx_{n_k} = -z_k + AP^+x_{n_k} + AP^-x_{n_k},$$

it follows that $((I-P)P_{n_k}Nx_{n_k}, (I-P)(x_{n_k} - x)) =$

$$= -(z_k, x_{1n_k} - x_1) + (AP^+x_{n_k}, P^+(x_{n_k} - x)) + (AP^-x_{n_k}, P^-(x_{n_k} - x))$$

$$\leq -(z_k, x_{1n_k} - x_1) + (AP^-x_{n_k}, P^-(x_{n_k} - x)) \to 0 \quad \text{as} \quad k \to \infty$$

Hence, by (2.39), we get $limsup\ (Nx_{n_k}, x_{n_k} - x) \leq 0$.

We have shown that $A - N = -(-A + N)$ is pseudo A-proper and therefore such is $-A + N$. To show the pseudo A-properness of $A + N$, we write $A + N = -(-A - N)$ and note that $-A$ is selfadjoint and

(2.41) $0\epsilon\sigma(-A)$ and $\sigma(-A) \cap (-\infty, 0) \neq \emptyset$ and consists of isolated eigenvalues having finite muliplicities.

Then $-\lambda_1$ is the largest negative eigenvalues of $-A$ and if $\{ E_\lambda \mid \lambda\epsilon R \}$ is the spectral resolution of A, we define $P^\pm$ by (2.24) and set $K = A^{-1}$. Then $KP^-$is positive semidefinite on $R(A)$ and KP^+ is compact on $R(A)$. Hence, using the same arguments as above we have that $-A - N$ is pseudo A-proper and such is $A + N$. ∎

For a smaller class of nonlinearities we shall now prove the A-properness of homotopies of the type in Theorem 2.5. Recall that $N : H \to H$ is said to be generalized pseudo-monotone if whenever $x_n \rightharpoonup x$ and $limsup$ $(Nx_n, x_n - x) \leq 0$ imply that $(Nx_n, x_n - x) \to 0$ and $Nx_n \rightharpoonup Nx$. It is of type (S+)

if $x_n \rightharpoonup x$ and $limsup(Nx_n, x_n - x) \leq 0$ imply that $x_n \to x$. If $x_n \rightharpoonup x$ implies that $limsup(Nx_n, x_n - x) \geq 0$, N is said to be of type (P). Note that if N is strongly monotone, i.e. $(Nx - Ny, x - y) \geq c \; \|x - y\|^2$ for all $x, y \in H$, then it is both of type (S) and (P).

For these classes of nonlinearities we have

PROPOSITION 2.3 ([Mi–5,6]) Let $A : D(A) \subset H \to H$ be a closed linear densely defined map with $R(A) = \ker A^{\perp}$ and either (2.36) or (2.37) holds. Let $N : H \to H$ be bounded and generalized pseudo monotone and $G : H \to H$ be bounded and of type (S) and (P). Then, for each $t\epsilon[0, 1), H_t = \pm\; A + (1-t)G + tN$ is A-proper w.r.t.$\Gamma = \{H_n, P_n\}$ for H with $P_nAx = Ax$ on H_n, and $\pm\; A + N$ is pseudo A-proper w.r.t.Γ.

Proof. We consider the case with $-A$ since the other case is treated similarly. Suppose first that (2.20) holds. Let $t\epsilon[0, 1)$ be fixed and $\{ x_{n_k} \in H_{n_k} \}$ be bounded and such that $-P_{n_k} H(t, x_{n_k}) \to f$ in H. We write $x_{n_k} = x_{on_k} + x_{1n_k}$ with $x_{0_{n_k}} \in H_0 = \ker A$ and $x_{1n_k} = \tilde{H} = R(A)$. It follows that

$$y_k \equiv Ax_{1n_k} - (1-t)(I-P)P_{n_k}Gx_{n_k} - t(I-P)P_{n_k}Nx_{n_k} \to f_1 \tag{2.42}$$

and

$$z_k \equiv (1-t)PP_{n_k}Gx_{n_k} + tPP_{n_k}Nx_{n_k} \to f_0. \tag{2.43}$$

Since G and N are bounded, so is $\{ Ax_{n_k} \}$ and we may assume that $x_{n_k} \rightharpoonup x_0 + x_1, Gx_{n_k} \rightharpoonup z, Nx_{n_k} \rightharpoonup y$ and $Ax_{n_k} \rightharpoonup u$. By the weak closedness of the graph of A, we have that $x\epsilon D(A)$ and $Ax = u$. Moreover, (2.42) and the compactness of A^{-1} imply that $x_{1n_k} \to x_1$. Hence, by (2.43).

$$(tNx_{n_k} + (1-t)Gx_{n_k}, x_{n_k} - x) \tag{2.44}$$

$$= (tP_{n_k}Nx_{n_k} + (1-t)P_{n_k}Gx_{n_k}, x_{n_k} - x) - (tNx_{n_k} + (1-t)Gx_{n_k}, (I - P_{n_k})x)$$

$$= (tPP_{n_k}Nx_{n_k} + (1-t)PP_{n_k}Gx_{n_k}, x_{n_k} - x)$$

$$+ (t(I-P)P_{n_k}Nx_{n_k} + (1-t)(I-P)P_{n_k}Gx_{n_k}, (I-P)(x_{n_k} - x))$$

$$- (tNx_{n_k} + (1-t)Gx_{n_k}, (I - P_{n_k})x) \to 0 \quad \text{as} \quad k \to \infty.$$

Hence,

$$limsup\,(tNx_{n_k}, x_{n_k} - x) \leq limsup(tNx_{n_k} + (1-t)Gx_{n_k}, x_{n_k} - x)$$
$$- (1-t)\,lim\,inf(Gx_{n_k}, x_{n_k} - x) \leq 0.$$

Let $t \neq 0$. Then $limsup(Nx_{n_k}, x_{n_k} - x) \leq 0$ and therefore $(Nx_{n_k}, x_{n_k} - x) \to 0$ and $Nx_{n_k} \rightharpoonup Nx$. Moreover, $limsup(Gx_{n_k}, x_{n_k} - x) = 0$ and therefore $x_{n_k} \to x$ by condition (S^+).
Next, let $t=0$ and note that by (2.44) $limsup(Gx_{n_k}, x_{n_k} - x) \leq 0$ and again $x_{n_k} \to x$. Moreover, $H(t,.) = f$ in either case since for each $h \in H$ and some $h_n \in H_n$ with $h_n \to h$,

$$(H(t,x), h) = \lim(P_{n_k}H(t, x_{n_k}), h_{n_k}) = (f, h).$$

Next, suppose that (2.37) holds. In view of the above discussion, we need only to show that for $t\in [0,1)$

$$limsup(tNx_{n_k} + (1-t)Gx_{n_k}, x_{n_k} - x) \leq 0 \tag{2.45}$$

where $\{x_{n_k}\}$ is as above. Using the decomposition (2.44), we see that (2.45) holds, since if we define $P^{\pm}$ as in the proof of Proposition 2.2, then

$$limsup\,(t(I-P)P_{n_k}Nx_{n_k} + (1-t)(I-P)P_{n_k}Gx_{n_k}, x_{1n_k} - x_1)$$
$$= \lim(-y_k, x_{1n_k} - x_1) + \lim\,\sup[(AP^+x_{n_k}, P^+(x_{n_k} - x)) +$$
$$(AP^-x_{n_k}, P^-(x_{n_k} - x))] \leq \lim\,\sup(AP^-x_{n_k}, P^-(x_{n_k} - x)) \leq 0.$$

Here we used that $(AP^+x, P^+x) \geq 0$ on $R(A)$. Finally, for $t = 0$, $H_1 = \pm A + N$ is pseudo A-proper w.r.t.Γ by Proposition 2.2 since a given bounded generalized pseudo monotone map is of type (M) (cf. [P–S]). ■

Remark 2.5 As G we can take P or $rI, r > 0$, or a strongly monotone map. Using the same type of arguments we can show that $\pm A + N + \alpha G$ is also A-proper w.r.t.Γ for each $\alpha > 0$ under the conditions of Proposition 2.3 (cf. [Mi–5]). As in [Mi–5,7] Proposition 2.3 is valid also for *g.p.m.(A)*,*P(A)* and *S(A)* type of maps when (2.36) or (2.37) holds.

Now, in view of Proposition 2.3, we have the following special case for Theorem 2.5.

THEOREM 2.7 Let $A : D(A) \subset H \to H$ satisfy (2.24)–(2.25) with $R(A) = \ker A^{\perp}$, and either (2.36) or (2.37) hold. Let $N : H \to H$ be a bounded monotone and asymptotically $\{B_1,B_2\}$–quasilinear map with $|N–K|$ small and suppose that some strongly monotone and selfadjoint map $C_0 : H \to H$ satisfies condition (2.26) for $\epsilon > 0$ sufficiently small. Then Eq. (1.1) is solvable for each $f \in H$.

To give some other special cases of Theorem 2.5, we need to discuss some other classes of (pseudo) A-proper maps. To that end, we need to introduce some concepts from the condensing mapping theory. For a bounded subset D of a Banach space X we define the ball-measure of noncompactness by

$$\chi(D) = inf\ \{r > 0 | D \subset \bigcup_{i=1}^{n} B(x_i, r), n \epsilon N, x_i \epsilon\ X\}$$

If Y is another Banach space, a map $F : D \subset X \to Y$ is said to be k-ball-contractive if $\chi(F(Q)) \leq k\chi(Q)$ for each $Q \subset D$; it is ball-condensing if $\chi(F(Q)) < \chi(Q)$ of each $Q \subset D$ with $\chi(Q) \neq 0$.

Let $A : D(A) \subset X \to Y$ be a closed linear densely defined map with $X_0 = ker\ A$ and $\widetilde{Y} = R(A)$ closed such that for some closed subspaces $\widetilde{X}$ and Y_0 of X and Y, respectively, $X = X_0 \oplus \widetilde{X}$ and $Y = Y_0 \oplus \widetilde{Y}$. Let $Q_0 : Y \to Y_0$ be a linear projection onto Y_0, $X_n = X_{0n} \oplus \widetilde{X}_{1n}$ and $Y_n = Y_{0n} \oplus Y_{1n}$ with $X_{0n} \subset X_0$, $X_{1n} \subset \widetilde{X}$, $X_{in} \subset X_{in+1}$, $i=0$, 1, and $Y_{on} \subset Y_0$, $Y_{1n} \subset \widetilde{Y}$.
If $Q_{0n} : Y_0 \to Y_{0n}$ and $Q_{1n} : \widetilde{Y} \to Y_{1n}$ are linear projections onto Y_{in}, $i = 0,1$, then $\Gamma = \{X_n, Y_n, Q_n = (Q_{0n}, Q_{1n})\}$ is a projection scheme for (X,Y) with $\delta_i = \max \| Q_{in} \| < \infty$, i=0,1.

Our next class of A-proper maps is given by

PROPOSITION 2.4 [Mi-5,7] Let the partial inverse $A^{-1} : \widetilde{Y} \to \widetilde{X}$ be compact, $N : X \to Y$ be bounded and such that for each fixed sequences $\{ x_{1n} \in X_{1n} \}$ and $\{ x_{on} \in X_{on} \}$ and some $c > 0$

$$\chi(\{Q_{0n}Q_0N(x_{0n} + x_{1n})\}) \geq c\chi(\{x_{0n}\}). \tag{2.46}$$

Let $F : X \to Y$ be continuous and k-ball-contractive with $k\delta_0 \| Q_0 \| < c$, or ball-condensing if $c = 1$. Supose that either N is continous or Y is reflective and N is demicontinuous. Then $\pm A + N + F : D(A) \subset X \to Y$ is A-proper w.r.t.Γ with $Q_{1n}Ax = Ax$ for $x \epsilon X_{1n}$.

Proof. Again we look at $A + N + F$ and the case $- A + N + F$ is studied similarly. Let $\{ x_{n_k} = x_{0n_k} + x_{1n_k} \epsilon D(A) \cap X_{n_k} \}$ be bounded and

$$y_{n_k} \equiv Q_{n_k}(A+N+F)x_{n_k} = Ax_{1n_k} + Q_{n_k}(N+F)x_{n_k} \to f. \tag{2.47}$$

Then

$$Ax_{1n_k} + Q_{1n_k}(I-Q_0)(N+F)x_{n_k} \to f_1 = (I-Q_0)f. \tag{2.48}$$

and

$$Q_{0n_k}\ Q_0\ Nx_{n_k} + Q_{0n_k}\ Q_0\ Fx_{n_k} \to f_0 = Q_0\ f. \tag{2.49}$$

Since N and F are bounded and A^{-1} is compact, (2.48) implies that we may assume that $x_{1n_k} \to x_1 \epsilon\ \widetilde{X}$. Moreover, by (2.46) and (2.49) we have

$$\begin{aligned} c\chi(\{x_{0n_k}\}) &\leq \chi(\{Q_{0n_k}Q_0Nx_{n_k}\}) = \chi(\{Q_{0n_k}Q_0Fx_{n_k}\}) \\ &\leq \delta_0 \| Q_0 \| \chi(\{Fx_{n_k}\}) \\ &\leq k\delta \| Q_0 \| \chi\{(x_{0n_k} + x_{1n_k)}\} = k\delta \| Q_0 \| \chi\{(x_{0n_k}\ )\}. \end{aligned}$$

Hence, $\chi(\{x_{0n_k}\}) = 0$ and therefore we may assume that $x_{0n_k} \to x_0$ and $x_{n_k} \to x =_0 + x_1$.

Next, for each $y \epsilon Y$, let $z_n \epsilon Y_n$ be such that $z_n \to y_0$. Then, by (2.47),

$$\begin{aligned} (Ax_{1n_k}, y) &= \lim(y_{n_k} - Q_{n_k}(N+F)x_{n_k}, z_{n_k}) = \lim(y_{n_k} - (N+F)x_{n_k}, Q^*{}_{n_k} z_{n_k}\} \\ &= (f - Nx - Fx, y) \end{aligned}$$

if $Y{=}Y^{**}$ and N is demicontinuous. Hence, $Ax_{1n_k} \to f - Nx - Fx$ and $x \in D(A)$ with $Ax + Nx + Fx = f$ by the weak closedness of A. Moreover, if N is continuous, then $Ax_{n_k} \to f - Nx - Fx$ and again $Ax + Nx + Fx = f$. Hence, $A + N + F$ is A-proper w.r.t.Γ in either case. ∎

Regarding condition (2.46) we have

LEMMA 2.7 Suppose that $N : X \to Y$ is demicontinuous and there is a constant $c > 0$ such that either one of the following conditions holds:

(2.50) N is strongly K-monotone, i.e. $(Nx - Ny, K(x-y)) \geq c \| x - y \|^2$ with $K : X \to Y^*$ such that $\| Kx \| \leq c_1 \| x \|$ and $Q_0^* Q_{0n}^* Kx = Kx$ for $x \in X_{0n}$

(2.51) For each $x_{0n}, z_{0n} \in X_{0n}$ and each $x_{1n} \in X_{1n}$,

$\| Q_{0n} \; Q_0 \; N(x_{0n}+x_{1n}) - Q_{0n} \; Q_0 \; N(z_{0n}+x_{1n}) \| \geq c \| x_{0n} - z_{0n} \|, n \geq 1.$

Then condition (2.46) holds.

Proof. It is clear that (2.50) implies (2.51). Let $\delta = \chi(\{Q_{0n}Q_0F(x_{0n}+x_{1n})\})$, $\epsilon > 0$ be fixed and cover $\{Q_{0n}Q_0F(x_{0n}+x_{1n})\}$ by finitely many balls $B(p_i, \delta+\epsilon)$, $p_i, \epsilon Y_0$, $i = 1, .., m$. Since $\bigcup Y_{0n}$ is dense in Y_0, for each i there exists $y_{ni} \in Y_{0n}$ such that $y_{ni} \to p_i$. For each n fixed, $Q_{0n}Q_0F(. \;, x_{1n}) : X_{0n} \to Y_{0n}$ is injective and continuous, and therefore the range of $Q_{0n}Q_0F(.,x_{1n})$ is open in Y_{0n} by the Brouwer invariance of domain theorem. Moreover, since its range is also closed in Y_{0n} by (2.51), it follows that $Q_{0n}Q_0F(. \;, x_{1n})(X_{0n}) = Y_{0n}$ for each n. Let $z_{ni} \in X_{0n}$ be such that $Q_{0n}Q_0F(z_{ni}+x_{1n}) = y_{ni}$ for each $i = 1, \ldots, m$ and each n. Choose $n \geq 1$ such that $\| y_{ni} - p_i \| < \epsilon$ for each $n \geq n_0$ and $1 \leq i \leq m$. Then, for each i fixed, let

$$N_i = \{ n | n \geq n_0, \; Q_{0n}Q_0F(x_{0n}+x_{in}) \in B(p_i, \delta+\epsilon \;) \} \; .$$

Now, for the smallest $\bar{n} \in N_i$, we have that $z_{\bar{n}i} \in X_{0n}$ for each $n \in N_i$ and

$$c \| x_{0n} - z_{\bar{n}i} \| \leq \| Q_{0n}Q_0F(x_{0n}+x_{1n}) - Q_{0n}Q_0F(z_{\bar{n}i}+x_{1n}) \| \leq$$

$$\| Q_{0n}Q_0F(x_{0n}+x_{1n}) - p_i \| + \| Q_{0n}Q_0F(z_{\bar{n}i}+x_{1n}) - p_i \| < \delta + \| p_i - y_{\bar{n}i} \| < \delta + 2\epsilon.$$

Hence, the set $\{cx_{0n}\}$ is covered by the balls $B(cz_{\bar{n}i}, \delta+2\epsilon) \subset X_0$, $1 \leq i \leq m$, and since $\epsilon > 0$ is arbitrary, we have that

$$c\chi(\{x_{0n}\}) \leq \chi(\{Q_{0n}Q_0F(x_{0n}+x_{1n})\}).$$

■

Recall that a map $N : X \to Y$ is of type (KS) for some $K : X \to Y^*$ if whenever $x_n \rightharpoonup x$ in X and limsup $(Nx_n, K(x_n - x)) \leq 0$, then $x_n \to x$. If $N : X \to Y$ is strongly K-monotone, i.e. $(Nx - Ny, K(x-y)) \geq c \| x - y \|^2$ for all x, y in X and some $c > 0$, it is of type (KS) provided that $\| Kx \| \leq c_1 \| x \|$ on X and $Q_0^* Q_{0n}^* Kx = Kx$ for x in X_{0n} .

Analyzing the proof of Proposition 2.4, we see that the following more general version of it is valid.

PROPOSITION 2.5 [Mi–5,7] Let $A^{-1}:\widetilde{Y}\to\widetilde{X}$ be compact and $N:X\to Y$ be bounded and either continuous or demicontinuous with $Y = Y^{**}$. Suppose that either one of the following conditions holds:
(2.52) N is of type (KS);
(2.53) If $\{x_{nk}=x_{on_k}+x_{1n_k}\in D(A)\cap X_{n_k}\}$ is bounded with $x_{1n_k}\to x_1\in\widetilde{X}$ and $Q_{0n_k}Q_0N(x_{0n_k}+x_{1n_k})\to Q_0f$, then $x_{0n_k}\to x_0$.

Then $\pm A+N:D(A)\subset X\to Y$ is A–proper w.r.t.Γ for (X,Y) with $Q_{1n}\ Ax = Ax$ on X_{1n}.

Finally, if A^{-1} in Proposition 2.5 is just continuous, we need then to impose stronger conditions on N. We have [Mi–7].

PROPOSITION 2.6 Let $A^{-1}:\widetilde{Y}\to\widetilde{X}$ be continuous and $N:X\to Y$ be continuous, k_1–ball–contractive and satisfy condition (2.46). Suppose that $F:X\to Y$ is continuous and k_2–ball–contractive with $(k_1+k_2)\delta_1\,\|I-Q_0\|\|A^{-1}\|+k_2\delta_0\,\|Q_0\|\,c^{-1}<1$. Then $\pm A+N+F:D(A)\subset X\to Y$ is A–proper w.r.t.Γ with $Q_{1n}Ax = Ax$ for $x\in X_{1n}$.

Proof. We look only at the "+" case, the other case being similar. Let $\{x_{n_k}=x_{0n_k}+x_{1n_k}\in D(A)\cap X_{n_k}\}$ be bounded and

$$y_{n_k}\equiv Q_{n_k}(A+N+F)x_{n_k}=Ax_{1n_k}+Q_{n_k}(N+F)x_{n_k}\to f.$$

Then, (2.48) – (2.49) hold and by the properties of N and F,

$$\chi(\{x_{1n_k}\})=\chi(\{A^{-1}Q_{1n_k}(I-Q_0)(N+F)x_{n_k}\})\le(k_1+k_2)\delta_1\,\|I-Q_0\|\|A^{-1}\|\,\chi(\{x_{n_k}\})$$

Hence, by (2.46) and (2.47),

$$c\chi(\{x_{0n_k}\})\le\chi(\{Q_{0n_k}Q_0Nx_{n_k}\})=\chi(\{Q_{0n_k}Q_0Fx_{n_k}\})\le k_2\delta_0\,\|Q_0\|\,\chi(\{x_{n_k}\}),$$

and therefore,

$$\chi(\{x_{n_k}\})\le\chi(\{x_{on_k}\})+\chi(\{x_{1n_k}\})\le$$

$$[(k_1+k_2)\delta_1\,\|I-Q_0\|\|A^{-1}\|+k_2/c\delta_0\,\|Q_0\|]\chi(\{x_{n_k}\}).$$

Thus, $\chi(\{x_{n_k}\})=0$ and we may assume that $x_{n_k}\to x$. Moreover, $Ax+Nx+Fx=f$ as in the proof of Proposition 2.4. ■

Now, to give another special case of Theorem 2.5, suppose that $N : H \to H$ has a symmetric weak Gateaux derivative $N'(x)$ on H such that

$$B^- \leq N'(x) \leq B^+ \text{ for } x \in H. \tag{2.54}$$

THEOREM 2.8 Let $A : D(A) \subset H \to H$ satisfy (2.24) - (2.25) with $R(A) = \ker A^\perp$, and (2.54) hold and $F : H \to H$ be continuous and k-ball-contracting with $k < \min\{\gamma_1, \gamma_2\}$ and small quasinorm $|F|$. Then the equation $Ax - Nx - Fx = f$ is solvable for each $f \in H$.

Proof. Let $x_0 \in H$ be fixed and $F_t = A - (1-t)N'(x_0) - tN$.
Set $K = P^+ - P^-$ with $P^\pm : H \to H^\pm$. We know that $K^2 = I$ and is a homeomorphism on H. For each $t \in [0, 1]$, it is easy to check that for each $x \in H$,

$$B_t^- \equiv tB^- + (1-t)N'(x_0) \leq (1-t)N'(x_0) + tN'(x) \leq tB^+ + (1-t)N'(x_0) \equiv B_t^+$$

and, by (2.24)–(2.25),

$$((A - B_t^-)x, x) \leq -\gamma_1 \| x \|^2 \quad \text{for} \quad x \in D(A) \cap X^-,$$

$$((A - B_t^+)x, x) \geq \gamma_2 \| x \|^2 \quad \text{for} \quad x \in D(A) \cap X^+.$$

Hence, for $x, y \in D(A)$ and $t \in [0, 1]$, $Nx - Ny = B(x-y)$, with $B = N'(y + a(x-y))$ for some $a \in (0, 1)$, and

$$\begin{aligned}(F_t(x) - F_t(y), K(x-y)) &= ([A - (1-t)N'(x_0) - tB]P^+(x-y), P^+(x-y)) \\ &\quad - ([A - (1-t)N'(x_0) - t\ B]P^-(x-y), P^-(x-y)) \\ &\geq ((A - B_t^+)P^+(x-y), P^+(x-y)) - ((A - B_t^-)P^-(x-y), P^-(x-y)) \\ &\geq \gamma\ (\| P^+(x-y) \|^2 + \| P^-(x-y) \|^2),\end{aligned}$$

where $\gamma = \min\{\gamma_1, \gamma_2\}$. Since for each $z \in H$

$$\| P^+z \pm P^-z \|^2 = \| P^+z \|^2 \pm 2(P^+z, P^-z) + \| P^-z \|^2$$

$$\leq \| P^+z \|^2 + 2 \| P^+z \| \ \| P^-z \| + \| P^-z \|^2 \leq 2(\ \| P^+z \|^2 + \| P^-z \|^2)$$

and $\| Kz \| \geq c \| z \|$ for some $c > 0$, we have that

$$(F_t(x) - F_t(y), K(x-y) \geq \gamma/2 \| K(x-y) \|^2 \geq c\gamma/2 \| x - y \|^2, x, \ y \in D(A).$$

Since F is k-ball contractive with $k < c\gamma/2$, Proposition 2.6 with A=0 implies that $H_t = F_t - tF$ is A-proper w.r.t.$\Gamma = \{ H_n, \ P_n \}$ for H with $P_nAx = Ax$ on H_n, for $t \in [0, \ 1]$.

Now, $N + F$ is asymptotically $\{B_1, B_2\}$-quasilinear with $B_1 = B^-, B_2 = B^+$ and $|N + F - K| = |F|$ small for each $x \in H$. Hence, the equation $Ax - Nx - Fx = f$ is solvable for each $f \in H$ by Theorem 2.5 with $C_0 = N'(x_0)$. ■

2.3 Uniqueness and Approximation-Solvability

In this section we shall study the unique approximation solvability of Eq. (1.1) and the rate of convergence of its approximate solutions. Since our maps are not Frechet differentiable, we cannot use the corresponding results of the author [Ph.D Thesis, Rutgers University, 1975] to study these problems. Instead, we shall use some extensions of these results to non-differentiable A-proper maps having multivalued derivative as developed in [Mi-10].

Let X and Y be Banach spaces with a projectionally complete scheme $\Gamma = \{X_n, P_n; Y_n, Q_n\}$, $U \subset X$ be an open subset and $T : \bar{U} \to Y$ be nonlinear.

Definition 2.3 A homogeneous map $B : X \to 2^Y$, with $B(x)$ convex for each $x \in X$, is a **multivalued derivative** of T at $x_0 \in U$ if there exists a map $R = R(x_0) : \bar{U} \setminus \{x_0\} \to 2^Y$ such that

$$Tx - Tx_0 \in B(x - x_0) + R(x - x_0) \quad \text{for } x \text{ near } x_0$$

and

$$\| y \| / \| x - x_0 \| \to 0 \quad \text{as} \quad x \to x_0 \quad \text{for each} \quad y \in R(x - x_0).$$

We shall need the following basic approximation-solvability and error estimate result for these maps.

THEOREM 2.9 [Mi-10] Let $T : \bar{U} \subset X \to Y$ be A-proper w.r.t.Γ, $x_0 \in U$ be an isolated solution of $Tx = f$ and $B : X \to 2^Y$ be a multivalued derivative of T at x_0 such that for some ϱ, $c_0 > 0$ and $n_0 \geq 1$

$$\| Q_n u \| \geq c_0 \| x \| \quad \text{for} \quad x \in B(0, \varrho) \cap X_n,\ u \in Bx, n \geq n_0. \tag{2.55}$$

Then the equation $Tx=f$ is strongly approximation-solvable in a ball $B(x_0, r)$ for some $r > 0$ (*i.e.*, $Q_n T x_n = Q_n f$ for some $x_n \in B(x_0, r) \cap X_n$ and all large n and $x_n \to x_0$). Moreover,

(a) If, in addition, there are some $c_1 > 0$ and $r_0 > 0$ such that $\| u \| \geq c_1 \| x \|$ for all $u \in Bx$ with $\| x \| \leq r_o$, then for any $\epsilon \in (0, c_o)$ the approximate solutions $x_n \in B(x_0, r) \cap X_n$ satisfy

$$\| x_n - x_0 \| \leq (c_0 - \epsilon)^{-1} \| Tx_n - f \| \quad \text{for} \quad n \geq n_1 \geq n_0. \tag{2.56}$$

(b) If $\| u \| \leq c_2 \| x \|$ for $u \in Bx$ with $\| x \| \leq r$ and some c_2 and if

(2.57) $Tx - Ty \in B(x - y) + R(x - y)$ whenever $x - y \in B(0, r)$ and $z / \| x - y \| \to 0$ as $x \to x_0$ and $y \to x_0$ for each $z \in R(x - y)$,

then the equation $Tx=f$ is uniquely approximation-solvable in B(x_0, r) and the unique approximate solutions $x_n \in B(x_0, r) \cap X_n$ satisfy

$$\| x_n - x_0 \| \leq k \| P_n x_0 - x_0 \| \leq c \; dist(x_0, X_n) \tag{2.58}$$

where a constant k depends on c_0, c_2, ϵ, $\delta = \max \| Q_n \|$, and $c = 2k\delta_1, \delta_1 = \max \| P_n \|$.

Let us now apply Theorem 2.9 to Eq. (1.1). Let $\delta = \max \| P_n \|$.

THEOREM 2.7 (cf. [Mi-10]) Let $A : D(A) \subset H \to H$ be a closed linear densely defined map and $C : H \to H$ be linear such that A–C is bijective and $\| (A-C)x \| \geq c_0 \| x \|$ for $x\epsilon D(A)$ and some c_0. Suppose that $N: H \to H$ is nonlinear, continuous and for some $a\epsilon(0, c_0/\delta)$

$$\| Nx - Ny - C(x-y) \| \leq a \| x - y \| \quad \text{for} \quad x, y\epsilon H \tag{2.59}$$

a) Then, for each $f\epsilon H$, Eq. (1.1) is uniquely approximation-solvable w.r.t. $\Gamma = \{H_n, P_n\}$ with $P_n(A-C)x = (A-C)x, x\epsilon H_n$, and the approximate solutions satisfy (2.56).

b) Let, there be in addition, a Hilbert space $V_1 \subset \ker A^{\perp} \subset V_1^*$, with each inclusion being continuous and dense, and a continuous extension A_1 of A wich is an isometric isomorphism of V_1 on to V_1^* and $\| (A_1 - C)x \|_V{}^* \geq c_1 \| x \|_V$ for $x\epsilon V = kerA \oplus V_1$ and some $c_1 > a$. Then the approximation solutions of Eq. (1.1) satisfy also (2.56).

Proof. Since $K = (N-C)(A-C)^{-1}: H \to H$ is a κ-contraction, $\kappa < 1$, $I-K$ is bijective by the contraction mapping principle, and therefore, such is also $A-N$. Moreover, $I-K$ is A-proper w.r.t.Γ and consequently so is $A-N$. It is easy to see that a map $B: H \to 2^H$ given by $Bx = \{y | \| y - Cx \| \leq a \| x \| \}$ is homogeneous and $Nx - Ny\epsilon B(x-y)$ for each x, $y\epsilon H$. Hence, $A-B$ is a multivalued derivative of $A-N$ at each $x\epsilon D(A)$ and satisfies condition (2.55) on H_n since for each $u\epsilon B(x), x\epsilon H_n$,

$$\| P_n(Ax-u) \| \geq \| (A-C)x \| - \| P_n(Cx-u) \| \geq (c_0 - \delta a) \| x \|$$

and $c_0 - \delta a > 0$. Similarly, using (2.59), we see that

$$\| A(x-y) - u \| \geq (c_0 - a) \| x - y \| \quad \text{for} \quad u\epsilon B(x-y), \quad x, y\epsilon D(A).$$

Since also $P_nK: H_n \to H_n$ is κ-contractive, $\kappa < 1$, we have that Eq. (1.1) is uniquely approximations solvable for each $f\epsilon H$. Moreover, by our discussion above and Theorem (2.9), the approximate solutions of Eq. (1.1) satisfy (2.56).

Next assume that conditions of b) hold. Then for $u\epsilon Bx$ with $\| x \|_V \leq r$ we have that

$$\| A_1 x - u \|_{V^*} \leq (\| A_1 \| + \| C \| + a) \| x \|_V . \tag{2.60}$$

Hence, by Theorem 2.9-(b) the rate of convergence of the approximate solutions x_n of (1.1) is given by

$$\| x_n - x_0 \|_V \leq k \| P_n x_0 - x_0 \|_V \leq c dist_V(x_0, H_n),$$

where a constant κ depends on c_1, a, $\| A_1 \|$, $\| C \|, \epsilon$ and δ , and $c = 2\kappa\delta$. ∎

In applications one usually takes $C = \lambda I$ for some $\lambda \notin \sigma(A)$. Regarding condition (2.59), we have

PROPOSITION 2.7 Let $A : D(A) \subset H \to H$ be selfadjoint, $\lambda = (\lambda_\kappa + \lambda_{\kappa+1})/2$ for some consecutive eigenvalues $\lambda = (\lambda_\kappa + \lambda_{\kappa+1})$ of A and $N : H \to H$ be nonlinear. Then condition (2.59) holds if either one of the following conditions is valid:

(2.61) N is a-strongly monotone and β-contractive with $\beta^2 < ka$ and $\lambda = \lambda_1/2 > 0$, where $\kappa = inf\{|\lambda_i| \mid \lambda_i \in \sigma(A)\backslash\{0\}\}$.

(2.62) N is Gateaux differentiable and for some $a, \beta \in R$ with $[a, \beta] \subset (\lambda_\kappa, \lambda_{\kappa+1})$:

$$a \| x - y \|^2 \leq (Nx - Ny,\ x - y) \leq \beta \| x - y \|^2 \text{ for } x, y \in H.$$

Proof. Let (2.61) hold. We have that $\beta < \kappa$ and

$$\| Nx - Ny - \lambda(x - y) \| \leq (\beta^2 + \lambda^2 - 2a\lambda)^{1/2} \| x - y \|^2 \text{ for } x, y \in H.$$

Since $\kappa \leq \lambda_1$ and $\beta^2 - \lambda_1 a < 0$, we have that $\beta^2 + \lambda^2 - 2a\lambda < \lambda_1^2/4$. Hence, (2.59) holds since

$$\| (A - \lambda I)^{-1} \|^{-1} = \min\{|\mu| \mid \mu \in \sigma(A - \lambda I)\}$$

$$= \min\{\lambda - \lambda_\kappa, \lambda_{\kappa+1} - \lambda\} = \lambda_1/2.$$

If (2.62) holds, then (2.59) is valid as shown in [Mi–10]. ∎

Remark 2.6 If $\sigma(A)$ is countable consisting of only eigenvalues and if the corresponding eignevectors form a basis in H, then it was proved in Smiley [Sm] that conditions of Theorem 2.10–b are valid. Moreover if (2.61) holds with $\kappa = inf\ \{\lambda_i | \lambda_i \in \sigma(A), \lambda_i < 0\}$, then the unique approximation solvability of $Ax - Nx = f$ was proved in [Sm] using the Liapunov–Schmidt alternative method and the obtained error estimates are of a different type. A numerical example can be also found in [Sm].

III. PERIODIC SOLUTIONS OF SEMILINEAR HYPERBOLIC EQUATIONS WITHOUT RESONANCE.

In this part we shall show how the abstract theory developed in Part 2 can be applied to the existence of weak T-periodic solutions of various hyperbolic equations without resonance. We begin by some preliminary considerations of abstract wave and telegraph equations and look in detail particular systems of such equations.

3.1 Abstract Semilinear Wave and Telegraph Equations.

Let H_1 be a real separable Hilbert space and $L : D(L) \subset H_1 \to H_1$ be a linear selfadjoint map with a compact resolvent. If $C(RxH_1, H_1)$ is the space of continuous functions from $RxH_1 \to H_1$, suppose that $F \in C(RxH_1, H_1)$ is T-periodic in t for some $T > 0$, i.e. $F(t+T, \cdot) = F(t, \cdot)$ for all $t \in R$. We are interested in the existence of T-periodic weak solutions for the abstract semilinear hyperbolic equations of the form

$$\ddot{u} + Lu + \sigma \dot{u} - F(t, u) = f(t), t \in R \tag{3.1}$$

where the dot denotes the time derivative and $\sigma \in R$. By a T-periodic solution of (3.1) we mean a function $u \in C^2(R, H_1) \cap L_2(R, D(L))$ such that

$$\ddot{u}(t) + Lu(t) + \sigma \dot{u}(t) - F(t, u(t)) = f(t), \quad u(t+T) = u(t)$$

for all $t \in R$, where $D(L)$ is endowed with the graph norm $\| \cdot \|_L$, i.e., $\| u \|_L^2 = \| Lu \|^2 + \| u \|^2$.

Let $H = L_2((0, T), H_1)$ and define a linear map $A_0 : D(A_0) \subset H \to H$ by

$$A_0 u(t) = \ddot{u}(t) + Lu(t) + \sigma \dot{u}(t), \text{ for all } t \in [0, T],$$

where

$$D(A_0) = \{u \in C^2([0, T], H_1) \cap L_2((0, T), D(L)) | u(0) = u(T), \dot{u}(0) = \dot{u}(T)\}.$$

A_0 is densely defined, and integration by parts shows that A_0 is symmetric, i.e., $A_0 \subset A_0^*$. Let $A = A_0^*$. Then we say that $u \in H$ is a **T-periodic weak solution** of (3.1) *iff*

$$u \in D(A) \text{ and } Au - F(t,u) = f(t) \text{ in } [0,T], \tag{3.2}$$

i.e., *iff*

$$\int_0^T (u(t), \ddot{v}(t) + Lv(t) + \sigma\dot{v}(t))dt - \int_0^T (v(t), F(t,u(t)))dt = \int_0^T (f(t), v(t))dt$$

for all $v \in D(A_0)$. Clearly, every T–periodic solution of (3.1) is a weak T–periodic solution.

Next, since L has a compact resolvent, there is an orthonormal basis $\{\psi_j | j \in J\}$ in H_1 and a sequence of its eigenvalues $\{\mu_j \in R | j \in J\}$ such that $|\mu| \to \infty$ (if $\dim H_1 = \infty$) Let

$$\psi_\kappa(t) = \begin{cases} c_\kappa \cos(\kappa\tau t) & \text{for} \quad \kappa \le 0 \\ c_\kappa jm(\kappa\tau t) & \text{for} \quad \kappa \ge 1 \end{cases}$$

for $t \in [0,T]$, where $\tau = 2\pi/T$ and $c_\kappa = \sqrt{2/T}$ if $\kappa \neq 0$ and $c_0 = 1/\sqrt{T}$. Define

$$\psi_{j\kappa}(t) = \psi_j \psi_\kappa(t), 0 \le t \le T.$$

Then $\{\psi_{j\kappa} | j \in J, \kappa \in Z\}$ is an orthonormal basis in H and $\psi_{j\kappa} \in D(A_0) \subset D(A)$. Hence, each function $u \in H$ can be expanded in the form

$$u(t) = \sum_{j,\kappa} a_{j\kappa}\psi_j e^{i\tau k t}, \quad a_{j,-k} = \bar{a}_{j,k}$$

and therefore

$$Au = \sum_{j,\kappa} a_{j\kappa}(\mu_j - \tau^2\kappa^2 + i\sigma\kappa)\psi_j e^{i\tau\kappa t}$$

for $u \in D(A)$, where

$$D(A) = \{u \in H | \sum_{j,\kappa} |a_{j\kappa}(\mu_j - \tau^2\kappa^2 + i\sigma\kappa|^2 < \infty\}$$

$$D(A) = \{u \in H | \sum_{j,\kappa} |a_{j\kappa}(\mu_j - \tau^2\kappa^2 + i\sigma\kappa|^2 < \infty\}$$

Using this representation of A, it is easy to see that A is a normal operator if $\sigma \neq 0$ and is symmetric if $\sigma = 0$. In the latter case

$$A^* = A_0^{**} \subset A_0^* = A \subset A_0^{**} = A^*$$

since A_0^{**} is the smallest closed extension of A_0, and therefore $A = A^*$, i.e., A is selfadjoint. Since the orthonormal basis $\{\psi_{j\kappa}\}$ of H consists of eigenfunctions of A, A has a pure point spectrum and, in particular, $\sigma(A) = \overline{\sigma_p(A)}$, where $\sigma_p(A) = \{\mu_j - \tau^2\kappa^2 | j \in J, \kappa \in Z\}$.

Now, if τ is a rational number, then 0 is not an accumulation point of σ(A), and therefore we can boundedly invert A on the orthogonal complement of κerA. (see the proof of Lemma 2.1). Hence, the range $R(A)$ of A is closed and the partial inverse $A^{-1} : R(A) \subset H \to H$ is continuous. However, depending on a choice of H_1, A^{-1} may or may not be compact. Actually, if σ (A) has nonzero accumulation points, then A^{-1} is not compact and we refer to the next section for a detailed discussion.

Define a nonlinear map $N : H \to H$ by $Nu = F(t, u)$. Then, (3.2) becomes a semilinear operator equation

$$Au - Nu = f \quad (u \in D(A), f \in H) \tag{3.3}$$

To apply the abstract theory of Part 2 to Eq. (3.3), we need the pseudo A–properness of $A - N$ with respect to a suitable scheme for H. This is so under some reasonable assumptions on F as discussed below. However, finding good concrete conditions on F that would imply the solvability of (3.3), and hence of (3.1), in this general setting is not an easy task. In what follows, we shall exhibit one set of such conditions. However, for special choices of H_1, we shall develop in Sections 3.2 and 3.3 and Part 4 a detailed existence theory for (3.1).

Let $F = F_1 + F_2$ satisfy

(3.4) F_1 and F_2 are Caratheodory functions and there are positive constants c and k_1, k_2 sufficiently small such that for a.e. $t \in (0,T)$ and all $x, y \in H_1$:

$$(F_1(t,x) - F(t,y), x-y) \geq c \| x-y \|^2_{H_1},$$

$$\| F_i(t,x) - F_i(t,y) \| \leq \kappa_i \| x-y \|_{H_1}, \quad i = 1,2 .$$

(3.5) There are positive constants a and b with a sufficiently small such that for some $\lambda \in (\lambda_i, \lambda_{i+1})$

$$\| F(t,x) - \lambda x \|_{H_1} \leq a \| x \|_{H_1} + b, \quad \text{for all } x \in H_1 .$$

THEOREM 3.1 Let (3.4) – (3.5) hold and τ be rational. Then there is a T–periodic weak solution of (3.1) with $\sigma = 0$ in H for each $f \in H$.

Proof. Set $N_i u = F_i(t,u)$ and $N = N_1 + N_2$. Since N_1 is c–strongly nonotone and k_1T–contractive and N_2 is k_2T–contractive with k_1 and k_2 sufficiently small, the map $A + N$ is A–proper w.r.t. a scheme Γ induced by the eigenfunctions of A by Proposition 2.6. Moreover, by (3.5),

$$\| Nu - \lambda u \| \leq a_1 \| u \| + b_1 \quad \text{for all } u \in H.$$

Hence, by Theorem 2.1 with X=H, Eq. (3.3) is solvable for each $f \in H$. ∎

3.2 Periodic Solutions of Semilinear Wave Equations Without Resonance.

In this section we shall study a special case of (3.1), with $\sigma = 0$, corresponding to a system of semilinear wave equations in $n \geq 1$ space dimensions. Let $Q \subset R^n$ be a bounded domain with smooth boundary and set $\Omega = (0,T)xQ$ and $H_1 = L_2(Q, R^m)$ for some $m \geq 1$. Then we can identify $H = L_2((0,T), L_2(Q, R^m))$ with $L_2(\Omega, R^m)$ with the inner product defined by

$$(u,v) = \int_0^T \int_Q (u(t,x), v(t,x)) dx dt$$

where $(u(t,x), v(t,x))$, $(t,x) \in \Omega$, is the inner product in R^m. Let L_1 be a linear selfadjoint elliptic operator in space variables $x \in R^n$ with coefficients independent of t such that the induced bilinear form $a(u, v)$ on the Sobolev space $W_2^1(Q, R^m)$ is continuous and symmetric. Suppose that V is a closed subspace of $W_2^1(Q, R)$, containing the test functions, such that $a(u, v)$ is semi-coercive on V, i.e. there are constants $a_1 > 0$ and $a_2 \geq 0$ such that

$$a(u, v) \geq a_1 \| u \|_{2,1}^2 - a_2 \| u \|^2_{L_2} \text{ for all } u \in V.$$

Define a linear map $L_0 : D(L_0) \subset L_2(Q, R^m) \to L_2(Q, R^m)$
by

$$(L_0 u, v) = a(u, v) \quad \text{for } each \ v \in V,$$

where

$$D(L_0) = \{u \in V | a(u, .) \text{ is continuous on V in the } L_2\text{-norm}\}.$$

It is well known that L_0 is selfadjoint and has a compact resolvent, since W_2^1 is compactly embedded in L_2. Next, define a selfadjoint map with compact resolvent $L : D(L) \subset H_1 = L_2(Q, R^m) \to H_1$ by:

$$D(L) = [D(L_0)]^m \text{ and } L = diag(L_0, \cdots, L_0) \ .$$

Let $F : RxQxR^m \to R^m$ be a Caratheodory function and consider the semilinear system of wave equations:

$$(3.6) \qquad \begin{cases} u_{tt} - L_1 u - F(t, x, u) = f(t, x) \\ u(t, .) \in V^m \end{cases}$$

where $f \in L_2(\Omega, R^m)$ is a T-periodic function in t variable and $\tau = 2\pi/T$ is rational.

By a *T*-**periodic weak solution** of the variational boundary value problem (3.6) for the semilinear system of wave equations we mean a solution of the nonlinear operator equation

$$Au - Nu = f, u \in D(A), \quad f \in H \tag{3.7}$$

where $\quad Au = \sum_{j,\kappa} u_{j\kappa}(\mu_j - \tau^2\kappa^2)\psi_j(x)e^{i\tau\kappa t}$ for

$$u \in D(A) = \{\ u = \sum_{j,\kappa} u_{j\kappa}\psi_j(x)e^{i\tau\kappa t} | \sum_{j,\kappa} |u_{j\kappa}(\mu_j - \tau^2\kappa^2)|^2 < \infty\}$$

and $Nu = F(t,x,u)$ for $u \in L_2(\Omega, R^m)$.

Regarding $F = F_1 + F_2$, we assume

(3.8) F_1 is a Caratheodory function such that for some $a_1 > 0, \kappa \in (0,1)$ and $h_1 \in L_2(\Omega, R)$ it satisfies

$$|F_1(t,x,y)| \le a_1|y|^\kappa + h_1(t,x) \text{ for } a.e.(t,x) \in \Omega, \text{ all } y \in R^m;$$

(3.9) F_2 is a Caratheodory function and there are $h_2 \in L_2(\Omega, R)$ and, for some consecutive eigenvalues $\lambda_i < \lambda_{i+1}$ of A, $\lambda \in (\lambda_i, \lambda_{i+1})$ and $0 < a_2 < \min\{\lambda - \lambda_i, \lambda_{i+1} - \lambda\}$ such that

$$|F_2(t,x,y) - \lambda y| \le a_2|y| + h_2(t,x), \text{ for } a.e.(t,x) \in \Omega,\ y \in R^m .$$

Our first nonresonance result for (3.6) is for non-monotone nonlinearities F and is an application of Theorem 2.1.

THEOREM 3.2 Let $Q = (0,\pi), V = \overset{\circ}{W}{}_2^1(Q,R), \mu_j \ge 0$ for each $j \in J$, and each nonzero eigenvalue of A is of finite multiplicity. Suppose that (3.8) – (3.9) hold. Then there is a T-periodic weak solution $u \in \overset{\circ}{W}{}_2^1$ of (3.6) for each $f \in L_2(\Omega, R^m)$.

Proof. Let $\{H_n\}$ be an increasing sequence of finite dimensional subspaces spanned by the eigenfunctions $\{\psi_j(x)e^{i\tau\kappa t}\}$ and $P_n : H \to H_n$ be the orthogonal projections onto H_n. Since the eigenfunctions are dense both in $\overset{\circ}{W}{}_2^1(\Omega, R^m)$ and $L_2(\Omega, R^m)$, $\Gamma = \{H_n, H_n, P_n\}$ in a projection scheme for $(\overset{\circ}{W}{}_2^1, L_2)$. Moreover, if $Nu = F(t,x,u)$, by the compactness of the embedding of W_2^1 into L_2 the map $A - N$: $D(A) \cap \overset{\circ}{W}{}_2^1(\Omega, R^m) \to L_2(\Omega, R^m)$ is pseudo A- proper w.r.t.Γ . In view Theorem 2.1, it remains to verify conditions (2.1)–(2.3).

First, we shall show that if $f \in \ker A^{\perp}$ and $Au = f$, then $u \in \overset{\circ}{W}{}_2^1$ and $\| u \|_{2,1} \leq c \| f \|$ for some $c > 0$ independent of f. Indeed, since

$$f = \sum_{j,k} f_{jk} \psi_j(x) e^{i\tau\kappa t} \quad \text{and}$$

$$Au = \sum_{j,k} u_{jk}(\mu_j - \tau^2\kappa^2)\psi_j(x)e^{i\tau k t}$$

then $Au = f$ implies that $u_{jk} = f_{jk}/(\mu_j - \tau^2 k^2)$ for $u_j \neq \tau^2 k^2$. Hence,

$$\| u \|_{L_2}^2 = \sum_{\mu_j \neq \tau^2 k^2} \frac{|f_{jk}|^2}{(\mu_j - \tau^2 k^2)^2} \leq \frac{1}{a^2} \sum_{j,k} |f_{jk}|^2 = a^{-2} \| f \|_{L_2}^2 \quad ,$$

where $a = \inf\{|\mu_j - \tau^2 k^2| \; |\mu_j \neq \tau^2 k^2\} > 0$. Moreover, since [Br–Ni–1] $u \in L_2$ belongs to $\overset{\circ}{W}{}_2^1$ *iff* $|u| = \sum_{j,k} (1 + |\lambda_j| + \tau^2 k^2)|u_{jk}|^2 < \infty$ and $|u|^{1/2}$ is equivalent to $\| u \|_{2,1}$, we have

$$\| u_t \|_{L_2}^2 + \sum_{i=1}^{n} \| u_{x_i} \|_{L_2}^2 \leq c \sum_i \sum_{\lambda j \neq \tau^2 k^2} |f_{jk}|^2 \frac{|\lambda_j| + \tau^2 k^2}{(\lambda_j - \tau^2 k^2)^2}$$

$$= \sum_i \Big(\sum_{\lambda_j > \tau^2 k^2} + \sum_{\lambda_j < \tau^2 k^2} \Big).$$

When $\lambda_j > \tau^2 k^2$, then $|\lambda_j| + \tau^2 k^2 < 2\lambda_j$ and

$$(\lambda_j - \tau^2 k^2)^2 = (\lambda_j^{1/2} - |\tau\kappa|)^2(\lambda_j^{1/2} + |\tau\kappa|)^2 \geq a^2\lambda_j.$$

If $\lambda_j < \tau^2\kappa^2$, then $|\lambda_j| + \tau^2\kappa_2 < 2\tau^2\kappa^2$ and $(\lambda_j - \tau^2\kappa^2)^2 \geq a^2(\lambda_j^{1/2} + |\tau\kappa|)^2 \geq a^2\tau^2\kappa^2$.

Hence,

$$\sum_{\lambda_j \neq \tau^2\kappa^2} |f_{jk}|^2 \frac{\lambda_j + \tau^2\kappa^2}{(\lambda_j - \tau^2\kappa^2)^2} \leq \frac{2}{a^2}\Big(\sum_{\lambda_j > \tau^2\kappa^2} |f_{jk}|^2 + \sum_{\lambda_j < \tau^2\kappa^2} |f_{jk}|^2 \Big)$$

and therefore $\| u \|_{2,1} \leq C \| f \|$. Then, by Lemma 2.1, condition (2.1) holds.

Next, set $N_i u = F_i(t,x,u)$ for $u \in L_2$, $i = 1,2$. Then, by the Minkowski and Hölder inequalities, (3.8) - (3.9) imply that for some constants c_1 and c_2

$$\| N_1 u \| \leq c_1 \| u \|^{\kappa} + \| h_1 \| \quad \text{and} \quad \| N_2 u \| \leq c_2 \| u \| + \| h_2 \| \quad \text{for} \quad u \in L_2 \ .$$

Since $\kappa < 1$, there are positive a, b and r such that $a < \min\{\lambda - \lambda_i, \ \lambda_{i+1} - \lambda\}$ and

$$\| Nu - \lambda u \| \leq a \| u \| + b \ \text{ for all } \ \| u \| \geq r.$$

Hence, (2.2) - (2.3) hold and Theorem 2.1 is applicable. ■

It is easy to see that condition (3.9) is implied by the following two conditions with $\lambda = (\lambda_i + \lambda_{i+1})/2$ and $a = (\lambda_{i+1} - \lambda_i)/2 - \epsilon$:

(3.10) There are constants $M > 0$ and $r > 0$ and $h \in L_2(\Omega, R)$ such that for each $1 \leq l \leq m$

$$|F_{2,l}(t,x,y)| \leq M|y_l| + h_2(t,x) \quad \text{for} \quad a.e.(t,x) \in \Omega, y \in R^m$$

(3.11) For $a.e.(t,x) \in \Omega, y = (9_1 \ldots, y_m) \in R^m$ with $|y_l| \geq r$:

$$\lambda_i + \epsilon \leq F_{2,l}(t,x,y)/_{y_l} \leq \lambda_{i+1} - \epsilon.$$

Hence, when $m = n = 1$, we have

COROLLARY 3.1 Let $F : Rx(0,\pi)xR \to R$ be a 2π-periodic in t Caratheodory function satisfying conditions (3.10) - (3.11). Then, for each $f \in L_2$, there is a 2π-periodic weak solution of $u \in \overset{\circ}{W}{}_2^1$ of

$$(3.12) \qquad \begin{cases} u_{tt} - u_{xx} - F(t,x,u) = f(t,x), & t \in R, x \in (0,\pi) \\ u(t,0) = u(t,\pi) = 0, & t \in R \\ u(t+2\pi, x) = u(t,x), & t \in R, x \in x \in (0,\pi). \end{cases}$$

Remark 3.1 The solvability of (3.12) for a dense set of f's in L_2 was proved by Hofer [Ho] under a global Lipschitz condition on F (cf. also [W]) and by Tanaka [Ta-1] without this condition. When F is monotone, Corollary 3.1 is due to Mawhin [Ma-2]

We continue our study of (3.6) when a nonlinear perturbation F satisfies asymptotic nonuniform nonresonance conditions with respect to two consecutive eigenvalues of the associated linear problem. These conditions are more general than (3.9) and (3.10) – (3.11) and our method of study requires a monotonicity conditions on F. When $m = n = 1$, this problem has been studied by Mawhin–Ward [Ma–Wa] for the wave equation (3.12).

Let A be the abstract realization of the linear problem associated with (3.6) and $\lambda_i < \lambda_{i+1}$ be two consecutive eigenvalues of A having finite multiplicities.

Let $F : \Omega x R^m \to R^m$ be a Caratheodory function such that for each $r > 0$ and $1 \le l \le m$ there are functions $a_l, \beta_l \in L_\infty(\Omega)$ and $h_r \in L_2(\Omega)$ such that

(3.13) $\quad |F(t,x,y)| \le h_r(t,x)$ for $a, e, (t,x) \in \Omega, |y| \le r,$

(3.14) $$a_l(t,x) \le \liminf_{|y_l| \to \infty} F_l(t,x,y)y_l^{-1} \le \limsup_{|y_l| \to \infty} F_l(t,x,y)y_l^{-1} \le \beta_l(t,x)$$

uniformly *a.e.* in $(t,x) \in \Omega$ and $(y_1, \cdots, y_{l-1}, y_{l+1} \cdots, y_m) \in R^{m-1}$, and

(3.15) $$\lambda_i \le a_l(t,x) \le \beta_l(t,x) \le \lambda_{i+1} \quad a.e. \text{ on } \Omega$$

with $\lambda_i < a_l(t,x)$ and $\beta_l(t,x) < \lambda_{i+1}$ on some sets of positive measure.

THEOREM 3.3 Let (3.13) – (3.15) hold and

(a) sign $\lambda_i F$ be monotone if $n = 1$, i.e., sign $\lambda_i(F(t,x,y) - F(t,x,z)) \cdot (y - z) \ge 0$ for $a.e.(t,x) \in \Omega$ and all $y, z \in R^m$;
(b) If $n > 1$, then $F = F_1 + F_2$ and for some positive constants c, κ_1, κ_2 with $(\kappa_1 + \kappa_2) \| A^{-1} \| < 1, \kappa_2 < c$;

(3.16) $\quad sgn\lambda_i(F_1(t,x,y) - F_1(t,x,z), y - z) \ge c|y - z|^2$ for $a.e.(t,x) \in \Omega, yz \in R^m$

(3.17) $\quad |F_i(t,x,y) - F_i(t,x,z)| \le \kappa_i|y - z|$ for $a.e.(t,x) \in \Omega, y, z \in R^m, i = 1, 2.$

Then there is a T-periodic weak solution $u \in L_2(\Omega, R^m)$ of (3.6) for each $f \in L_2(\Omega, R^m)$.

Proof. We have that either $0 < \lambda_i < \lambda_{i+1}$ or $\lambda_i < \lambda_{i+1} < 0$. We may assume that $\lambda_i > 0$, for otherwise instead of the corresponding operator equation

$$Au - Nu = f, u \in D(A), f \in H = L_2(\Omega, R^m) \tag{3.18}$$

where $Nu = F(t, x, u)$, we can consider the equivalent equation

$$A_1 u - N_1 u = -f$$

with $A_1 = -A, N_1 = -N, \sigma(A_1) = \{\ldots < 0 < -\lambda_{i+1} < \lambda_i < \ldots\}$. Then, setting

$$\alpha_{1l} = -\beta_l, \beta_{1l} = -\alpha_l, \widetilde{F} = -F,$$

we see that conditions (3.14), (3.15) and (3.17) hold with $\alpha_l, \beta_l, F, \lambda_i$ and λ_{i+1} replaced respectively by $\alpha_{1l}, \beta_{1l}, \widetilde{F}, -\lambda_i$ and $-\lambda_{i+1}$ and the function $sign(-\lambda_i)\widetilde{F}_1 = sign\lambda_i F_1$ is monotone if $n = 1$, or satisfies (3.16) if $n > 1$. Hence, we can assume that $\lambda_i > 0$ and therefore $N : H \to H$ is monotone when $n = 1$ and N is c-strongly monotone and N_j are k_j-contractive, $j = 1, 2$, when $n > 1$.

Next, we shall show that N is a bounded asymptotically $\{B_1, B_2\}$-quasilinear map with $B_1 = C_1 - \epsilon I$ and $B_2 = C_2 + \epsilon I$ for some $\epsilon > 0$, where C_1 and C_2 are mxm diagonal matrices with the diagonal entries $\alpha_1(t,x), \ldots, \alpha_m(t,x)$ and $\beta_1(t,x), \ldots, \beta_m(t,x)$ respectively. By (3.14), for $\epsilon > 0$ there is an $r > 0$ such that for each $1 \le l \le m$, for $a.e.(t,x) \in \Omega$ and all $y = (y_1, \ldots, y_l, \ldots, y_m) \in R^m$ with $|y_l| \ge r$:

$$\alpha_l(t,x) - \epsilon \le F_l(t,x,y) y_l^{-1} \le \beta_l(t,x) + \epsilon. \tag{3.19}$$

Hence, by (3.13),

$$|F_l(t,x,y)| \le (\lambda_{i+1} + \epsilon)|y_l| + h_r(t,x) \text{ for } a.e.(t,x) \in \Omega, y \in R^m,$$

and therefore N is continuous and bounded in H. For each $1 \le l \le m$, define a function $G_l : \Omega x R^m \to R$ by (cf. [Ma–Wa])

$$G_l(t,x,y) = \begin{cases} y_l^{-1}F_l(t,x,y), & \text{if } |y_l| \geq r \\ r^{-1}F_l(t,x,y_1\ldots,y_{l-1},r,y_{l+1},\ldots,y_m)\frac{y_l}{r} + (1-\frac{y_l}{r})a_l(t,x), & \text{if } 0 \leq y_l \leq r \\ r^{-1}F_l(t,x,y_1\ldots,y_{l-1},-r,y_{l+1},\ldots,y_m)\frac{y_l}{r} + (1+\frac{y_l}{r})a_l(t,x), & \text{if } -r \leq y_l \leq 0, \end{cases}$$

Using (3.19), it is easy to check that for $1 \leq l \leq m$

$$a_l(t,x) - \epsilon \leq G_l(t,x,y) \leq \beta_l(t,x) + \epsilon \qquad \text{for } a.e.(t,x) \in \Omega, y \in R^m$$

Moreover, $H_l(t,x,y) = F_l(t,x,y) - G_l(t,x,y)y_l$ is a Caratheodory function on $\Omega x R^m$ for $1 \leq l \leq m$ and

$$|H_l(t,x,y)| \leq 2h_r(t,x) \quad \text{for } a.e.(t,x) \in \Omega, y \in R^m.$$

Let $G(t,x,y)$ be the mxm diagonal matrix with the diagonal entries $G_1(t,x,y)$, ... $G_m(t,x,y)$, $(t,x) \in \Omega, y \in R^m$.

Similarly, let $H(t,x_1\ y)$ be the $m\ x\ m$ diagonal matrix with the diagonal entries $H_1(t,x,y)$, ..., $H_m(t,x,y)$.

Now, for each $u \in H$, define $B(u): H \to H$ by $B(u)v = G(t,x,u)v$ and $Mu = H(t,x,u)$. Then, $Nu = B(u)u + Mu$ on H, and it is easy to check that $B(u)$ is selfadjoint and $B_1 \leq B(u) \leq B_2$ for each $u \in H$. Since $\| Mu \| \leq 2 \| h_r \|$ for each $u \in H$, it follows that N is asyomptotically $\{B_1,B_2\}$-quasilinear. Moreover, by Lemma 2.5, conditions (2.24) – (2.25) hold for any selfadjoint maps $B^{\pm}$ with $C_1 \leq B^-$ and $B^+ \leq C_2$, where H^+ (resp., H^-) is a subspace of H spanned by the eigefunctions of A corresponding to the eigenvalues $\lambda_k \leq \lambda_i$ (resp.,$\lambda_k \geq \lambda_{i+1}$).

Finally, define the selfadjoint map $C_0: H \to H$ by $C_0u: a(t,x)u$, where $a(t,x)$ is the mxm diagonal matrix with the diagonal entries $a_1(t,x),\ldots,a_m(t,x)$, for $(t,x) \in \Omega$.

Since each $a_l(t,x) \geq \lambda_i > 0$ $a.e.$ on Ω, it follows that C_0 is continuous and a_i-strongly monotone. Hence, $H(t,x) = A - (1-t)C_0 - tN$ is A-proper $w.r.t.\Gamma = \{H_n, P_n\}$ with $P_nAx = Ax$ on H_n for each $t \in [0,1)$ if $n \geq 1$, and H_1 is pseudo A-proper if $n = 1$ and is A-proper if $n > 1$ by Propositions 2.3 and 2.6. Choosing $\epsilon > 0$ as in Lemma 2.5, we see that the conclusion of the theorem follows from Theorems 2.7 and 2.8. ■

Remark 3.2 When $n = m=1$, Theorem 3.3 with the Dirichlet boundary conditions for the semilinear wave equation (3.12) was proved by Mawhin–Ward [Ma–Wa] using rather different arguments based on the compactness of A^{-1} and the coincidence degree theory of Mawhin. Hence, their method is not applicable when $n > 1$ since A^{-1} is not compact due to the presence of nonzero eigenvalues of infinite multiplicity.

Remark 3.3 In view of Remark 2.4, Theorem 3.3 still remains valid if $\lambda_i = 0$ (resp., $\lambda_{i+1} = 0$) provided each α_l (resp., β_l) is a constant.

Our final result in this section deals with (3.6) when $D_yF(t,x,y)$ exists and a nonresonance condition is such that it requires the eigenvalues of the matrix $D_yF(t,x,y)$ to lie in possibly distinct gaps of the spectrum of A. Such problems with $n = 1$ have been studied in [La, A, Br–Li, Ma–3] and [Am–2] when $n \geq 1$. Using the abstract results on asymptotically $\{B_1, B_2\}$ –quasilinear maps from Part 2, we shall now give an extension of a result of Amann [Am–2].

Denote by $\mathcal{L}s(R^m)$ the set of alll symmetric linear endomorphisms of R^m, which we can (and will) identify canonically with the symmetric $m \times m$ matrices. Let $F = F_1 + F_2$ and impose the following conditions:

(3.20) $F_1 : \Omega x R^m \to R^m$ is a Caratheodory function such that $F_1(t,x,\cdot) \in C^1(R^m, R^m)$, with a symmetric derivative $D_yF_1(t,x,y) \in \mathcal{L}_s(R^m)$ for $a.e.(t,x) \in \Omega$ and all $y \in R^m$.

(3.21) There exist matrices $b^+, b^- \in \mathcal{L}\ s(R^m)$ such that

$$b^- \leq D_yF_1(t,x,y) \leq b^+ \quad \text{for} \quad a.e.(t,x) \in \Omega, y \in R^m.$$

(3.22) $F_2 : \Omega x R^m \to R^m$ is a Caratheodory function and there are a sufficiently small $a > 0$ and $h \in L_2(\Omega, R)$ such that

$$|F_2(t,x,y)| \leq a|y| + h(t,x) \quad \text{for} \quad a.e.(t,x) \in \Omega, y \in R^m.$$

(3.23) There is a $k > 0$ sufficiently small such that

$$|F_2(t,x,y) - F_2(t,x,z)| \leq \kappa|y-z| \quad \text{for} \quad a.e.(t,x) \in \Omega, y, z \in R^m$$

We note that (3.21) implies that $F_1(t,x,.)$ is the gradient of some function $\phi(t,x,\cdot) : R^m \to R$ for $a.e.(t,x) \in \Omega$. For every $b \in \mathcal{L}_s(R^m)$, we define a continuous linear map $B \in L(H,H)$, where $H = L_2(\Omega, R^m)$, by

$$(Bu)(t,x) = bu(t,x) \text{ for } all\ u \in H \text{ and } a.e.(t,x) \in \Omega .$$

Then B is said to be the constant mutiplication operator induced by b, and $\sigma(B) = \sigma(b)$ is the point spectrum. Let $B^\pm$ denote the constant multiplication operators induced by $b^\pm$, and let $\lambda_1^\pm \le \ldots \le \lambda_m^\pm$ be the eigenvalues of $b^\pm$, where each eigenvalue is repeated according to its multiplicity.

Let $\{e_i^\pm | i = 1, \ldots, m\}$ be an orthonormal basis for R^m such that $e_i^\pm$ is an eigenvector to the eigenvalue $\lambda_i^\pm$ of $b^\pm$. Then the spectral resolution of $b^\pm$ is

$$b^\pm = \sum_{i=1}^{m} \lambda_i^\pm (e_i^\pm, \cdot) e_i^\pm$$

and therefore

$$(b^\pm y, y) = \sum_{i=1}^{m} \lambda_i^\pm (e_i^\pm, y)^2 \text{ for } all\ y \in R^m$$

where (,) is the Euclidean inner product. Hence, replacing $b^\pm$ by

$$b_\epsilon^\pm = \sum_{i=1}^{m} (\lambda_i^\pm \pm \epsilon_i)(e_i^\pm, \cdot) e_i^\pm \in \mathcal{L}_s(R^m)$$

where $\epsilon_i \ge 0$ are sufficiently small, we may assume that the eigenvalues $\lambda_i^\pm$ of $b^\pm$ are pairwise disjoint.

Let $A : D(A) \subset H \to H$ be the abstract realization of the linear problem associated with (3.6) and $\sigma(A)$ be its resolvent set. Then A is selfadjoint. Assume the nonresonance condition of the form:

(3.24) A commutes with $B^\pm$ and $\bigcup_{i=1}^{m} [\lambda_i^-, \lambda_i^+] \subset \varrho(A)$

THEOREM 3.4 Let (3.20) - (3.24) hold and b^+ and b^- commute. Then there is a T-periodic weak solution of (3.6) for each $f \in L_2(\Omega, R^m)$.

Proof. It is well known that (3.20) - (3.21) imply that $N_1 u = F_1(t, x, u)$ maps H into H and has a Gateaux derivative N_1' on H such that

$$B^- \leq N_1'(u) <= B^+ \text{ for } all\ u \in H.$$

Note that $B^\pm$ has the spectral resolution

$$B^\pm = \sum_{i=1}^{m} \lambda_i^\pm P_i^\pm,$$

where $P_i^\pm$ is the orthogonal projection onto the eigenspace $ker(B^\pm - \lambda_i^\pm I)$. It is clear that $P_i^\pm$ is the constant multiplication operator induced by the projection. $P_i^\pm = (e_i^\pm, \cdot)e_i^\pm : R^m \to Re_i^\pm$. It is easy to see that, after a possible renumeration of the eigenvalues of $b^\pm$, we can assume that $P_i^+ = P_i^-$, for $1 \leq i \leq m$.

Now, in view of our discussion above, Lemma 2.5 and Theorem 2.8, it remains only to show that $N_2 u = F_2(t, x, u)$ is a k-ball contractive map on H with a sufficiently small quasinorm $|N_2|$. But, this follows from the fact that N_2 is k-contractive on H by (3.23) and (3.22). ■

When $F_2 \equiv 0$, Theorem 3.4 was proved by Amann [Am-2] in a rather different manner using the monotone operator theory. In this case one has also the unique solvability of (3.6) (cf. [Am-2]). On the other hand, Amann's result is a vast extension of the earlier results of Lazer [La], Ahmad [A], Brown-Lin [Br-Li] and Mawhin [Ma-3].

3.3 Periodic Solutions of Semilinear Telegraph Equations Without Resonance.

In this section we shall first study a special case of (3.1) with $\sigma \neq 0$, corresponding to a nonresonant system of semilinear telegraph equations in $n \geq 1$ space dimensions satisfying the Dirichlet conditions on ∂Q, where $Q \subset R^n$ is a bounded domain with smooth boundary. More precisely, let $\Omega = (0, T)xQ$ with $T = 2\pi/\tau, \tau$ a rational number and let L_1 be as in Section 3.2, $F : RxQxR^m \to R^m$ be a T-peridic in t Caratheodory function and consider the semilinear system of telegraph equations

$$(3.25) \qquad \begin{cases} u_{tt} - L_1 u + \delta u_t - F(t,x,u) = f(t,x) \ , (t,x) \in \Omega \\ u(t,x) = 0 \ \text{ for } \ (t,x) \in [0,T]x\partial Q \end{cases}$$

where $\sigma \neq 0$ and $f \in H = L_2(\Omega, R^m)$ is a T-periodic function in t.

By a **T-periodic weak solution** of the variational boundary value problem (3.25) for the semilinear system of telegraph equations we mean a solution of the nonlinear operator equation.

$$Au - Nu = f, \quad u \in D(A) \subset H, f \in H \tag{3.26}$$

where $$Au = \sum_{j,k} u_{jk}(\mu_j - \tau^2 k^2 + i\sigma\tau k)\psi_j(x)e^{i\tau kt}, \quad u_{j,-k} = \bar{u}_{j,k},$$

for $$u \in D(A) = \{u = \sum_{j,k} u_{jk}\psi_j(x)e^{i\tau kt} \,\Big|\, \sum_{j,k} |u_{jk}(\mu_j - \tau^2 k^2 + i\sigma\tau k)|^2 < \infty\},$$

and $Nu = F(t,x,u)$ for $u \in H$.

In the second part of the section we shall study systems of semilinear telegraph equations with a nonlimear dissipation term without resonance. Such problems have been studied by many authors (cf. [Br–Ni–1, Ma–1, Ra, V] and the literature in there).

We begin by looking first at (3.25) when the null space $N(A) \neq \{0\}$. If $\mu_0 = 0$ is an eigenvalue of L (a variational extension of L_1 as defined in section 3.1) of multiplicity d, then $N(A)$ is d-dimensional and is spanned by the eigenfunctions $\psi_1(x), \ldots, \psi_d(x)$ of L corresponding to $\mu_0 = 0$. Then for a real number $\lambda \notin \sigma(L)$ and $f \in H$ we have

$$A_\lambda^{-1} f \equiv (A - \lambda I)^{-1} f = \sum_{j,k} \frac{f_{jk}}{\mu_j - \lambda - \tau^2 k^2 + i\sigma\tau k} \psi_j(x) e^{i\tau kt},$$

$$A_\lambda^{-1}f \in \overset{\circ}{W}{}_2^1(\Omega, R^m) \text{ and (cf. [Br–Ni–1])}$$

$$\| A_\lambda^{-1}f \|_{2,1}^2 = \sum_{j,k} |f_{jk}|^2 \frac{1 + |\lambda_j| + \tau^2\kappa^2}{|\mu_j - \lambda - \tau^2\kappa^2 + i\sigma\tau\kappa|^2}$$

$$\leq c_1^2(L, \lambda, \tau) \, \| f \|^2 .$$

Hence, $A_\lambda^{-1} : L_2 \to D(A) \cap \overset{\circ}{W}{}_2^1$ is continuous and therefore compact as a map in L_2. Similarly, we have

$$\| A_\lambda^{-1}f \| \leq c(L, \lambda, \tau) \, \| f \| \quad \text{for } f \in L_2, \tag{3.27}$$

and it is easy to see that $c(L, \lambda, \tau) < 1/dist(\lambda, \sum)$, where $\sum = \{\mu_j - \tau^2\kappa^2 | j \in J, \kappa \in Z\}$.

Our nonresonance result in this case is an easy application of a version of Theorem 2.2.

THEOREM 3.5 Let $F = F_1 + F_2, F_1, F_2 : \Omega x R^m \to R^m$ be Caratheodory functions such that F_1 satisfies (3.8) and for some $\lambda \notin \sigma(L), h_2 \in L_2(\Omega, R)$ and $\epsilon > 0$:

$$|F_2(t, x, y) - \lambda y| \leq \frac{1}{c(L, \lambda, \tau) + \epsilon} |y| + h_2(t, x) \tag{3.28}$$

for $a.e. (t, x) \in \Omega$ and $y \in R^m$. Then there is a T-periodic weak solution $u \in \overset{\circ}{W}{}_2^1(\Omega, R^m)$ of (3.25) for each $f \in L_2$.

Proof. Let $H_n = linspan\{\psi_j(x)e^{i\tau\kappa t} \mid |j|, |\kappa| \leq n\}$ and $P_n : H \to H_n$ be the orthogonal projection for each n. Since $A_\lambda^{-1} : H \to H_n$ is compact and $Nu = F_1(t, x, u) + F_2(t, x, u)$ is a bounded map on H, it is easy to show that the map $T = A - N : D(A) \subset H \to H$ is A-proper w.r.t.$\Gamma = \{H_n, P_n\}$. Moveover, by (3.8) and (3.28),

$$\| Nu - \lambda u \| \leq (c(L, \lambda, \tau) + \epsilon)^{-1} \, \| u \| \quad \text{for} \quad \| u \| \geq R \tag{3.29}$$

Then, using the homotopy $H(t, u) = (A - \lambda I)u - t(N - \lambda I)u$ and (3.27) and 3.29), we get as in the proof of Theorem 2.2 that the equation $Au - Nu = f$ is solvable in H

for each $f \in H$ (cf. [Mi-6]). Since $A_\lambda; H \to \overset{\circ}{W}{}_2^1$, it follows that each solution belongs to $\overset{\circ}{W}{}_2^1(\Omega, R^m)$. ■

Remark 3.4 Since $c(L, \lambda, \tau) < 1/dist(\lambda, \sum)$, condition (3.28) is implied by (3.9) or (3.10) - (3.11) with $\lambda_i, \lambda_{i+1} \in \sum$.

Next, as a preparatory step for studying systems of semilinear telegraph equations with nonlinear damping term, we consider first the existence of T-periodic weak solutions of the system

$$(3.30) \qquad \begin{cases} u_{tt} + L_1 u + F(t, x, u_t) = f(t, x), (t, x) \in \Omega \\ u(t, x) = 0 \quad \text{for} \quad (t, x) \in [0, T]x\partial Q. \end{cases}$$

As for (3.25), we use the same approach for (3.30) with the perturbing term of the form σu_t due to the type of the nonlinearity F. Hence, we need an estimate on the weak solutions of the linear problem.

$$(3.31) \qquad \begin{cases} u_{tt} + L_1 u + \sigma u_t = f, \\ u(t, x) = 0 \quad on \quad [0, T]x\partial Q \end{cases}$$

for $f \in L_2$. For definiteness suppose $\sigma > 0$ and note that [Br–Ni–1, Ra] any T-periodic weak solution u of (3.31) belongs to $\overset{\circ}{W}{}_2^1(\Omega, R^m)$ and there is a constant $c > 0$ such that $\| u \|_{2,1} \leq c \| f \|$. When $N(A) = N(L) = \{0\}$, we shall now get an explicit form of c.

LEMMA 3.1 Let $\sigma(L) = \{\mu_j > 0, j = 1, 2, ...\}$ and $f \in L_2(\Omega, R^m)$. If $u \in \overset{\circ}{W}{}_2^1(\Omega, R^m)$ is the unique solution of (3.31), then

$$\| u \|_{2,1} \leq [\sigma^{-1} + 2^{-1} \cdot (\mu_1^{-1/2} + (\mu_1^{-1} + 4\sigma^{-2})^{1/2})] \| f \| \tag{3.32}$$

Proof. Write $A = A_1 + \sigma C$, where $Cu = u_t$ and

$$A_1 u = \sum_{j,k} u_{jk}(\mu_j - \tau^2 \kappa^2)\psi_j(x) e^{i\tau\kappa t}, \quad u_{j,-\kappa} = \bar{u}_{jk}$$

Then, for $u \in D(A_1) \cap \overset{\circ}{W}{}_2^1$ we have

$$(A_1 u, Cu) = (\sum_{j,k} u_{j,k}(\mu_j - \tau^2 k^2)\psi_j e^{i\tau\kappa t}, \sum_{j,k} i\kappa\; u_{jk}\psi_j e^{i\tau\kappa t})$$

$$= - \sum_{j,k} i|u_{jk}|^2 \kappa(\mu_j - \tau^2\kappa^2) = 0$$

since $u_{j,-k} = \overline{u_{jk}}$ and for each j fixed and $n \geq 1$

$$\sum_{|k| \leq n} \kappa |u_{jk}|^2 (\mu_j - \tau^2\kappa^2) = 0.$$

Similarly, we get $(Cu, u) = 0$ and

$$(A_1 u, u) = \sum_{j,k} \mu_j |u_{jk}|^2 - \sum_{j,k} \tau^2\kappa^2 |u_{jk}|^2 = (Lu, u) - \| Cu \|^2 .$$

Moreover, $(Lu, u) \geq \mu_1 \| u \|^2$ for each $u \in \overset{\circ}{W}{}_2^1(\Omega, R^m)$. Now, taking the inner product of (3.31) with Cu and then with u, we get respectively

$$\| Cu \| \leq \frac{1}{\sigma} \| f \| \tag{3.33}$$

and

$$(Lu, u) \leq \frac{1}{\sigma^2} \| f \|^2 + \| f \| \, \|u\| \leq \frac{1}{\sigma^2} \| f \| + \mu_1^{-1/2} (Lu, u)^{1/2} \| f \| .$$

Hence,

$$(Lu, u)^{1/2} \leq \frac{1}{2} [\mu_1^{-1/2} + (\mu_1^{-1} + 4\sigma^{-2})^{1/2}] \, \| f \| . \tag{3.34}$$

Since $\| u \|_{2,1} = (Lu, u)^{1/2} + \| Cu \|$ for $u \in \overset{\circ}{W}{}_2^1$, we get (3.32). ■

Remark 3.5 When $m = 1$ and $L_1 u = -\Delta_n$, Lemma 3.1 was proved by Prodi [Pr] using different arguments.

For (3.30) we have

THEOREM 3.6 Let $\sigma(L) = \{\mu_j > 0 | j = 1, 2, \ldots\}$, $F_1, F_2 : \Omega x R^m \to R^m$ be Caratheodory functions such that F_1 satisfies (3.8) and for some $\sigma > 0, h_2 \in L_2(\Omega, R)$ and $\epsilon > 0$:

$$|F_2(t,x,y) - \sigma y| \leq \frac{1}{c(\sigma) + \epsilon}|y| + h_2(t,x), \tag{3.35}$$

for $a.e.(t,x) \in \Omega$, *all* $y \in R^m$ and $c(\sigma) = 1/\sigma + (\mu_1^{-1/2} + (\mu_1^{-1} + 4\sigma^{-2})^{1/2})/2$. Then there is a T–periodic weak solution $u \in \overset{\circ}{W}{}_2^1(\Omega, R^m)$ of (3.30) for each $f \in L_2$. The solution is unique if $F(t,x,\cdot)$ is strictly monotone, i.e., $(F(t,x,y) - F(t,x,z)) \cdot (y - z) > 0$ for $a.e.(t,x) \in \Omega$ and all $y \neq z$ *in* R^m.

Proof. As before, the existence of T–periodic weak solution of (3.30) reduces to the solvability of the operator equation

$$Au + Nu = f \qquad (u \in D(A), f \in L_2) \tag{3.36}$$

where A is the abstract realization of the linear problem and $Nu = F(t,x,u)$. Set $Cu = u_t$ for $u \in \overset{\circ}{W}{}_2^1(\Omega, R^m)$. By (3.8) and (3.35) we get

$$\| Nu - \sigma Cu \| \leq (c(\sigma) + \epsilon)^{-1} \| u \|_{2,1} + b$$

for $u \in \overset{\circ}{W}{}_2^1$ and some b > 0. Moreover, by Lemma 3.1,

$$\| u \|_{2,1} \leq c(\sigma) \| Au + \sigma Cu \| \quad \text{for } u \in \overset{\circ}{W}{}_2^1.$$

Since the scheme $\Gamma = \{H_n, H_n, P_n\}$, with $H_n = linspan\{\psi_j(x)e^{i\tau\kappa t} | j, |\kappa| \leq n\}$ and $P_n : H :\to H_n$ is the orthogonal projection, is such that $P_n(A + \sigma C)u = (A + \sigma C)u$ for $u \in H_n$, the solvability of (3.36) follows from Theorem 2.2. Its unique solvability follows from the strict C–monotonicity of N, i.e., $Nu - Nv, C(u - v)) > 0$ for all $u \neq v$ *in* $\overset{\circ}{W}{}_2^1$ and $(Au, Cu) = 0$ by Remark 2.3.

Let us now look at a special case studied by Prodi [Pr] when $m = 1$.

Corollary 3.2 Let $\sigma(L) = \{\mu_j > 0 | j = 1, 2, \ldots\}$ and $F : \Omega x R^m \to R^m$ be a Caratheodory function such that $F(t, x, 0) = 0$ *on* Ω and for some $0 < a \leq \beta$ and each $1 \leq l \leq m$:

$$(3.36) \qquad a \leq (F_l(t, x, y) - F_l(t, x, z))(y_l - z_l)^{-1} \leq \beta$$

for $a.e.(t, x) \in \Omega$ and *all* $y \neq z$ *in* R^m. Then there is a unique T-periodic weak solution of (3.30) in $\overset{\circ}{W}{}_2^1$ for each $f \in L_2$.

Proof. In view of Theorem 3.6, we need only to show that condition (3.35) holds for F for some $\sigma > 0$ and $h \in L_2$. If $a = \beta$ then taking $\sigma = a$, we see that (3.35) trivially holds since $c(\sigma) > 0$ and $F(t, x, y) - \sigma y = 0$ for $(t, x) \in \Omega$ and $y \in R^m$. Hence, we may assume that $a < \beta$. Let $\sigma \in (0, \beta - a)$ to be selected later on. Then for each $(t, x) \in \Omega$ and $y \in R^m$ there is a $\theta(t, x) \in [a, \beta]$ such that,

$$F(t, x, y) - \sigma y = (\beta - \theta(t, x) - \sigma)y$$

and therefore,

$$|F(t, x, y) - \sigma y| \leq (\beta - a - \sigma)|y|$$

It remains to show that $(c(\sigma) + \epsilon)(\beta - a - \sigma) \leq 1$ for some $\epsilon > 0$. Define a function $g(\sigma) = (\beta - a - \sigma)c(\sigma)$ for $\sigma > 0$. Then, setting $a = \mu_1^{1/2}$, and using the fact that $(a + b)^p \leq a^p + b^p$ for all $a, b \geq 0$ and $0 < p \leq 1$, we get

$$g(\sigma) = \frac{2a + \sigma + (4a^2 + \sigma^2)^{1/2}}{2a\sigma}(\beta - a - \sigma) \leq \frac{2a + \sigma}{a\sigma}(\beta - a - \sigma).$$

Since the function $g_1(\sigma) = (2a + \sigma)(a\sigma)^{-1}(\beta - a - \sigma)$is montonically decreasing in $(0, \infty)$, $g_1(\sigma) \to \infty$ *as* $\sigma \to 0, g_1(\sigma) \to -\infty$ as $\sigma \to \infty$ and $g_1(\beta - a) = 0$, we can select $\sigma \in (0, \beta - a)$ such that $(c(\sigma) + \epsilon)(\beta - a - \sigma) \leq 1$ for some $\epsilon > 0$. ∎

Remark 3.6 When $m = 1$ and $L = -\Delta_n$ Corollary 3.2 was proved by Prodi [Pr] using quite different arguments based on the convengence of a certain iteration procedure. Actually in the setting of Corollary 3.2, we could show that the problem (3.30) is uniquely approximation–solvable and could find the rate of convengence of the approximate solution (using Theorem 2.10)

Finally, we shall consider the existence of T–periodic weak solutions of semilinear systems of telegraph equation with nonlinear damping:

$$\begin{cases} u_{tt} + L_1 u + F(t,x,u_t) - G(t,x,u) = f(t,x) \;\; on \;\; \Omega \\ u(t,x) = 0 \;\; on \;\; [0,T]x\partial Q. \end{cases} \tag{3.37}$$

Regarding F we assume condition (3.35) and

(3.38) $G : \Omega x R^m \to R^m$ is a Caratheodory function such that for $a.e.(t,x) \in \Omega$,

all $y \in R^m$ and some $s > 0$ and $g \in L_2$

$$|G(t,x,y)| \leq s|y| + g(t,x).$$

It follows from (3.35) that there are a constant $a > 0$ and functions $h_1 \in L_2(\Omega, R^+)$ and $s(x) \in L_\infty(Q, R^m)$ such that

$$|F(t,x,y) - s(x)y| \leq a|y| + h_1(t,x) \tag{3.39}$$

for $a.e.(t,x) \in \Omega,$ all $y \in R^m.$

By (3.38), there are $b \geq 0$, $\phi \in L_2(\Omega, R^+)$ and a Caratheodory function $r : Q x R^m \to R^m$ having a linear growth such that

$$|G(t,x,y) - r(x,y)| \leq b|y| + \phi(t,x) \quad on \;\; \Omega x R^m. \tag{3.40}$$

If G does not depend on t, b could be zero. Moreover, there are $d > 0$ and $\psi \in L_2(\Omega, R^+)$ such that

$$(G(t,x,y), y) \leq d|y|^2 + \psi(t,x)|y| \quad \text{on} \;\; \Omega x R^m. \tag{3.41}$$

We suppose all that $F(t,x,.)$ is coercive on R^m, i.e. for some $m > 0$:

$$(F(t,x,y), y) \geq m|y|^2 \;\; \text{for} \;\; a.e.(t,x) \in \Omega, y \in R^m. \tag{3.42}$$

Let a*, b* and d* be the least upper bounds on a, b and d for which (3.39) – (3.41) hold. Suppose

$$m^2(\lambda_1 - d^*) - ma^*b^* - b^{*2} > 0. \tag{3.43}$$

We have for (3.37)

THEOREM 3.7 Let $\sigma(L) = \{\lambda_j > 0 | j = 1, 2, \ldots\}$, $F : \Omega x R^m \to R^m$ satisfy (3.35) and (3.42) and G satisfy (3.38). Suppose that (3.43) holds. Then there is a T-periodic weak solution $u \in \overset{\circ}{W}{}_2^1$ of (3.37) for each $f \in L_2$.

Proof. Let $X = \overset{\circ}{W}{}_2^1(\Omega, R^m)$, $H = L_2(\Omega, R^m)$ and A be the linear map induced by the associated linear problem to (3.38). Define the maps

$Cu = u_t$, $Mu = F(t, x, u_t)$ and $Su = s(x)Cu$ on X, and $Nu = G(t, x, u)$ and $Ru = r(x, u)$ on H. Choose a, b and d so that (3.39) – (3.41) hold for some suitable functions $r(x, y)$, $s(x)$, $\phi(t, x)$ and $\psi(t, x)$ and that, by (3.43),

$$m^2(\lambda_1 - d) - mab - b^2 > 0.$$

It is well known that $N : H \to H$ is continuous as is $M_1 : H \to H$, $M_1 u = F(t, x, u)$. Since $\| Cu \| \leq \| u \|_{2,1}$, $C : X \to H$ is continuous and since $C(H_n) \subset H_n$, $C : H_n \to H_n$ is continuous. But, $Mu = M_1(Cu)$ and therefore $P_n M : H \to H_n$ is continuous. Moreover, for each $u \in H_n$,

$$\| P_n(A + M)u \| \geq \| (A + \sigma C)u \| - \| (M - \sigma C)u \|$$

$$\geq \left(\frac{1}{c(\sigma)} - \frac{1}{c(\sigma) + \epsilon}\right) \| u \|_{2,1} - b = c_0 \| u \|_{2,1} - b$$

with $c_0 > 0$, and consequently condition (2.23) is satisfied. By (3.35), from the proofs of Theorem 3.6 and 2.2, we have that

$$\deg(P_n(A + M), B(0, R) \cap H_n, 0) \neq 0, n \geq n_0 \text{ for } all\ R > 0 \text{ large.}$$

Hence, in view of our discussion above, all other conditions of Corollary 2.1 are satisfied, and consequently the problem (3.37) has a T-periodic weak solution $u \in \overset{\circ}{W}{}_2^1$ for each $f \in L_2$. ■

Remark 3.7 When $m = 1$, Theorem 3.7 was proved by Prodi [Pr] using the Leray-Schauder degree theory under the additional assumptions that F is strongly monotone and β- contractive on R^m with $0 < \alpha \leq \beta$ (i.e. F satisfies (3.36)). Moreover, under (3.36) he has also shown that one can allow that G depends also on $(u_{x_1}, \ldots, u_{x_n})$ provided it satisfies the Lipschitz condition in u and $grad_x u$.

IV. SEMILINEAR EQUATIONS AT RESONANCE

In this section, we shall study Eq. (1.1) when dim $\ker A = \infty$ and the linearity N interacts in some sense with the spectrum of A. We begin by allowing resonance conditions of the antipodes type. Then. we show that some Landesman - Lazer type of of resonance conditions, extensively studied when A is a Fredholm map of nonnegative index, also guarantee the solvability of Eq. (1.1) in our setting. We require that $A+N : D(A) \cap X \subset X \to H$ is pseudo A-proper which is so, in particular, if X is compactly embedded in H. Hence, no monotonicity of N is required.

In the second part of Section 4.1, we discuss an extension by the author of the basic result of Brezis - Nirenberg [Br–Ni–1] to pseudo A-proper maps $A + N$. Finally, a perturbation method for hyperbolic problems at resonance is studied based on closeness of $A + N$. As discussed in Section 2 and also in this section, $A + N$ is pseudo A-proper or closed for some classes of maps N even if A^{-1} is not compact. Hence, the obtained results are suitable also for applications to resonance problems for systems of semilinear hyperbolic equations in more than one space variable.

4.1 Resonance Problems - Direct Methods

Throughout the section we assume that $(X, \|\cdot\|_0)$ is a Banach space continuously and densely embedded in a Hilbert space H and $A : D(A) \subset H \to H$ is a closed linear densely defined map with infinite dimensional kernel H_0 and closed range $\widetilde{H} = H_0^{\perp}$. Then $H = H_0 \oplus \widetilde{H}$ and $X = X_0 \oplus \widetilde{X}$, where $X_0 = H_0 \cap X$ and $\widetilde{X} = \widetilde{H} \cap X$ are closed subspaces of X. If $\Gamma = \{H_n, H_n, P_n\}$ is a projection scheme for (X, H) with $\| P_n \| = 1$, then $H_n = H_{0n} \oplus H_{1n}$, with $H_{0_n} = H_0 \cap H_n = X_0 \cap H_n$ and $H_{1n} = \widetilde{H} \cap H_n = \widetilde{X} \cap H_n$, and $P_n = (P_{0n}, P_{1n})$ where $P_{0n} : H_0 \to H_{on}$ and $P_{1n} : \widetilde{H} \to H_{1n}$ are orthogonal projections. Let $P : H \to H_0$ be the orthogonal projection.

For $x \in X, x = x_0 + x_1$ with $x_0 \in X_0, x_1 \in \widetilde{X}$ and define a new norm $\| x \|_1 = \| x_0 + x_1 \|_1 = \max\{ \| x_0 \|, \| x_1 \| \}$. Regarding A we assume

(4.1) There is a constant $c > 0$ such that if $Ax = f$ for $f \in R(A)$, then $x \in X$ and $\| x \|_0 \leq c \| f \|$.

We impose the following antipodes and growth conditions on $N: H \to H$.

(4.2) For a given $f_0 \in \widetilde{H}$ there are constants $M \geq 0, K > 0, \varrho_0 \geq 0$ and $n_0 \geq 1$ such that for each $r > K, x_1 \in B_X(0,r) \cap H_{1n}$, $x_0 \in \partial B_X(0,1) \cap H_{0n}, \varrho \geq rM + \varrho_0$ and $n \geq n_0$:

$$P_{on}PN(\varrho x_0 + x_1) - P_{on}f_0 \neq \mu P_{on}PN(-\varrho x_0 - x_1) - \mu P_{0n}f_0 \quad \text{for } \mu \in [0,1].$$

(4.3) $S = (I-P)N$ is quasibounded, i.e.

$$|S| = \limsup_{\|x\|_1 \to \infty} \frac{\|Sx\|}{\|x\|_1} < \infty \quad \text{and} \quad c|S| \max\{1, M\} < 1.$$

For such maps we have the following result announced in [Mi–5].

THEOREM 4.1 Suppose that conditions (4.1) – (4.3) hold and $A + N : D(A) \cap X \subset X \to H$ is pseudo A-proper w.r.t. $\Gamma = \{H_n, H_n, P_n\}$ with $P_nAx = Ax$ *on* H_n. Then Eq. (1.1) is solvable for each $f \in f_0 \oplus \widetilde{H}$.

Proof. Let $f = f_0 + f_1 \in f_0 \oplus \widetilde{H}$ be fixed and $R_0 > K$ such that $\|Sx\| < |S|\,\|x\|_1$ for each $\|x\|_1 \geq R_0$. Then there is an $R \geq R_0$ such that for each $r \geq R$ independent of n and $n \geq n_0$

$$P_{1n}(I-P)(Ax - Nx - f) \neq \mu P_{1n}(I-P)(-Ax - N(-x) - f) \tag{4.4}$$

for each $x = x_0 + x_1 \in H_{0n} \oplus H_{1n}$ with $\|x_0\|_0 < rM + \varrho_0, \|x_1\|_0 = r$ and $\mu \in [0,1]$. If not, then there would exist $x_{n_k} = x_{0n_k} + x_{1n} \in H_{0n_k} \oplus H_{1n_k}$ and $\mu_k \in [0,1]$ such that $\|x_{0n_k}\|_0 < r_kM + \varrho_0, \|x_{1n_k}\|_0 = r_k \to \infty$ and

$$P_{1n_k}(I-P)(Ax_{n_k} - Nx_{n_k} - f) = \mu_k P_{1n_k}(I-P)(-Ax_{n_k} - N(-x_{n_k}) - f)$$

for each k. Then

$$(1+\mu_k)P_{1n_k}Ax_{1n_k} = P_{1n_k}(I-P)(Nx_{n_k} - \mu_k N(-x_{n_k})) + (1-\mu_k)P_{1n_k}f_1$$

and therefore

$$(1+\mu_k)\|x_{1n_k}\|_0 \leq c(1+\mu_k)\|Ax_{1n_k}\| < c(1+\mu_k)|S|\,\|x_{n_k}\|_1 + c(1-\mu_k)\|f_1\|.$$

Hence, since $\| x_{n_k} \|_1 = \max\{ \| x_{0n_k} \|_0, \| x_{1n_k} \|_0 \}$, we get

$$1 < c|S|\{1, M + \varrho_0/r_k\} + c(1 - \mu_k)(1 + \mu_k)^{-1} \| f_1 \| / r_k.$$

Passing to the limit as $K \to \infty$, we obtain $1 \leq c|S| \max\{1, M\}$ in contradiction to (4.3). Hence, there is an $R \geq R_0$ such that (4.4) holds for each $n \geq n_0$ and the indicated $x \in H_n$.

Now, let $r \geq R$ be fixed and for each $n \geq n_0$, define

$$D_n \equiv D_n(r) = \{x = x_0 + x_1 \in H_{0n} \oplus H_{1n} | \, \| x_0 \|_0 \leq rM + \varrho_0, \| x_1 \|_0 \leq r\}$$

and the homotopy $H_n : [0, 1]x\bar{D}_n \to H_n$ by

$$H_n(t, x) = Ax + \frac{1}{1+t} P_n(Nx - f) - \frac{1}{1+t} P_n(N(-x) - f).$$

Then,

$$H_n(t, x) \neq 0 \quad \text{for} \quad x \in \partial D_n, t \in [0, 1], n \geq n_0. \tag{4.5}$$

Indeed, if $x \in \partial\bar{D}_n$ is such that $\| x_0 \|_0 < rM + \varrho_0$, then $\| x_1 \|_0 = r$ and (4.5) holds by (4.4). On the other hand, if $x \in \partial\bar{D}_n$ is such that $\| x_1 \|_0 < r$, then $\| x_0 \|_0 = rM + \varrho_0$ and

$$P_{0n}PN(x_0 + x_1) - P_{0n}f_0 \neq tP_{0n}PN(-x_0 - x_1) - tP_{0n}f_0$$

by (4.2) and therefore (4.5) holds again. Set $N_f x = Nx - f$. Since D_n is symmetric relative to the origin and $H_n(1, \cdot)$ is an odd map, the Borsuk theorem implies

$$\deg(A + P_n N_f, D_n, 0) = \deg(P_n H_1, D_n, 0) \neq 0, n \geq n_0.$$

Hence, $Ax_n + P_n N x_n = P_n f$ for some $x_n \in D_n$ and all $n \geq n_0$ and, by the pseudo A-properness of $A + N$, there is an $x \in X \cap D(A)$ such that $Ax + Nx = f$. ∎

In view of Proposition 2.1, we have the following special case.

COROLLARY 4.1 Suppose that conditions (4.1) – (4.3) hold and that X is compactly embedded in H. If N is bounded and demicontinuous, then Eq. (1.1) is solvable for each $f \in f_0 \oplus \widetilde{H}$.

We note that this result does not require monotonicity of N as is usually the case in dealing with Eq. (1.1) with $\dim\ker A = \infty$.

If $S = (I - P)N$ has a a sublinear growth, then condition (4.3) can be relaxed as in the following result.

THEOREM 4.2 [Mi–5] Let (4.1) hold and

(4.6) There are constants a, b and $\gamma \in (0, 1]$ such that

$$\| Nx \| \leq a + b \| x \|_1^\gamma \quad \text{for all} \quad \| x \|_1 \quad \text{large,}$$

(4.7) For a given $f_0 \in H_0$ there are constants $M \geq 0, n_0 \geq 1$ and $K > 0$ such that for each $r > K, x_1 \in B_X(0, r) \cap H_{1n}$, each $\varrho \geq r\varrho(r)$ for some $\varrho(r) \geq M$ and $x_0 \in \partial B_X(0, 1) \cap H_{on}$

$$P_{0n}PN(\varrho x_0 + \varrho^\gamma x_1) - P_{0n}f_0 \neq \mu P_{0n}PN(-\varrho x_0 - \varrho^\gamma x_1) - \mu P_{0n}f_0, \text{ for } \mu \in [0, 1], n \geq n_0.$$

Suppose that $cb < 1$ if $\gamma = 1$ and either X is compactly embedded in H or $A + N : D(A) \cap X \subset X \to H$ is pseudo A-proper w.r.t.$\Gamma = \{H_n, H_n, P_n\}$ with $P_n Ax = Ax$ on H_n. Then Eq. (1.1) is solvable for each $f \in f_0 \oplus \widetilde{H}$.

The proof of Theorem 4.2 can be given using similar arguments (cf. [Mi–4]).

COROLLARY 4.2 Theorems 4.1 and 4.2 are valid if conditions (4.2) and (4.7), respectively, are replaced by either one of the following inequalities

(4.8) $(PN(\varrho x_0 + \varrho^\gamma x_1), x_0) < (f_0, x_0)$, or

(4.9) $(PN(\varrho x_0 + \varrho^\gamma x_1), x_0) > (f_0, x_0)$

for all x_0 and x_1 as in (4.2) with $\gamma = 0$ or in (4.7), respectively.

Proof. It suffices to note that either (4.8) or (4.9) imply (4.2) and (4.7), respectively. ■

Remark 4.1 Condition (4.2) in Theorem 4.1 can be relaxed to $P_{0n}PN(\varrho x_0 + x_1) \neq tP_{0n}f_0$ for $t \in [0, 1]$ and x_0 and x_1 as in (4.2) provided we require that $\deg(P_{1n}A + P_{0n}PNP, D(r) \cap H_n, 0) \neq 0$ for all large n. This can be proved using the arguments similar to those in the proof of Theorem 5 in [Mi–4]. A similar extension of Theorem 6 in [Mi–4] can be also proved in the case of $\dim\ker A = \infty$. We omit the proofs.

Next, we shall study Eq. (1.1) under the following resonance conditions of Landesman – Lazer type introduced in [Fig, Fit] when N has a sublinear growth and A is Fredholm of index zero. Let $f_0 \in \widetilde{H}$ be given.

(4.10) $N : H \to H$ is asymptotically zero, i.e. $\| Nx \| / \| x \| \to 0$ *as* $\| x \| \to 0$, and either (*i*) $\lim inf(Nx_n, y) < (f_0, y)$ or (*ii*) $\lim sup(Nx_n, y) > (f_0, y)$ whenever $\{x_n\} \subset X$ is such that $\| x_n \|_0 \to \infty$ and $x_n / \| x_n \|_0 \to y \in H_0$ $\{0\}$ in H.

(4.11) N has a sublinear growth, i.e. $\| Nx \| \leq a + b \| x \|^k$ for $x \in H$ and some positive a, b and $k \in (0, 1)$, and either (*i*) $\lim inf(N(\varrho_n u_n + \varrho_n^{\gamma} v_n), x_0) < (f_o, x_o)$ or (*ii*) $\lim sup(N(\varrho_n u_n + \varrho_n^{\gamma} v_n), x_0) > (f_o, x_o)$ whenever $\varrho_n \to \infty$, $\{v_n\} \subset \widetilde{X}$ is bounded in X and $\{u_n\} \subset H_0$ is such that $u_n \rightharpoonup x_0 \neq 0$ in X.

Regarding the linear part, we assume

(4.12) Let 0 be an isolated point of the spectrum $\sigma(A)$ and for each $\lambda \in R$, with $|\lambda|$ small, there is a constant $c(\lambda) > 0$ such that $x \in X$ and $\| x \|_0 \leq c(\lambda) \| f \|$ whenever $(A + \lambda I)x = f$ for $f \in H$ if $\lambda \neq 0$ and $f \in R(A)$ otherwise.

The following result has been proved by the author [Mi–7] using only the Brouwer degree theory.

THEOREM 4.3 Let X be compactly embedded in H, (4.12) and either (4.10) or (4.11) hold. If $A + N : D(A) \cap X \subset X \to H$ is pseudo A-proper $w.r.t.\Gamma = \{H_n, H_n, P_n\}$ for (X, H) with $P_n Ax = Ax$ *on* H_n, then Eq. (1.1) is solvable for each $f \in f_0 \oplus \widetilde{H}$.

Remark 4.2 Using similar arguments as in [Mi–4,7], it can be shown that Theorem 4.3 is valid if (4.10) or (4.11) are replaced by other types of resonance conditions studied in [Mi–4,5] when A is Fredholm of a nonnegative index.

We continue our exposition by studying Eq. (1.1) under a resonance condition introduced by Brezis-Nirenberg [Br–Ni–1], but assuming only that $A + N$ is pseudo A-proper in H.

Assume that $A: D(A) \subset H \to H$ satisifies:

(4.13) There are positive constants $a_\pm$ and a_0 such that

(*i*) $\quad -a_+^{-1} \| Ax \|^2 \leq (Ax, x) \leq a_-^{-1} \| Ax \|^2$ for $x \in D(A)$,

(*ii*) $\quad \| x \| \leq a_0 \| Ax \|$ for $x \in \widetilde{H} = \ker A^\perp$

Let *Int(D)* denote the interior of *D* and *convD* be the convex hull of *D*. We have [Mi–7]

THEOREM 4.4 Let a linear closed map $A: D(A) \subset H \to H$ satisfy (4.13) and $N: H \to H$ be such that $\pm A + N: D(A) \subset H \to H$ is pseudo *A*–proper $w.r.t. \Gamma = \{H_n, H_n, P_n\}$ with $P_n Ax = Ax$ *on* H_n. Suppose that

(4.14) There are $\gamma < a_\pm$ and $\tau < a_0^{-2}(\gamma^{-1} - a_\pm^{-1})$ such that for every $y \in H$ and every $\delta > 0$ there exist $c_i(y), i = 1, 2,$ and $k(\delta)$ such that for each $x \in H$

$$(Nx - Ny, x) \geq \gamma^{-1} \| Nx \|^2 - c_1(y) \| Nx \| - \tau \| x_1 \| - c_2(y)(\delta \| x_0 \| + k(\delta)).$$

Then $Int(R(A) + convR(N)) \subset R(\pm A + N)$. Moreover, if *N* is onto, so is $\pm A + N$.

Now, condition (4.14) implies that $\limsup_{\| x \| \to \infty} \| Nx \| / \| x \| \leq \gamma$, i.e. *N* is a quasibounded map with the quasinorm $|N| \leq \gamma$. As pointed out in [Br–Ni–1] , if *N* is a potential and quasibounded, then a condition of type (4.14) holds. Moreover, if a is the smallest positive constant such that $(Ax, x) \geq -a^{-1} \| Ax \|^2$ on $D(A)$ and *A* is selfadjoint, then $Ax = \lambda x$ implies that $\lambda \geq -a$ and $-a$ is an eigenvalue of *A*. Hence, roughly speaking, the conditions $0 < \gamma < a$ and $\limsup_{\| x \| \to \infty} \| Nx \| / \| x \| \leq \gamma$, mean that the nonlinearity *N* asymptotically stays away from the nonzero eigenvalues of *A*.

Note that $a_\pm$ in (4.13) exist since, by the boundedness of $A^{-1}: R(A) \subset H \to H$, $\| Ax_1 \| \geq a \| x_1 \|$ for some $a > 0$ and all $x_1 \in D(A) \cap R(A)$, and therefore $(Ax, x) = (Ax, x_1) \geq -\| Ax \| \| x_1 \| \geq -a^{-1} \| Ax \|^2$ for all $x \in D(A)$. Here we used the fact that $R(A) = N(A)^\perp$. If *A* is sefadjoint and $a < \infty$ is the smallest positive constant such that $(Ax, x) \geq -a^{-1} \| Ax \|^2$ *on* $D(A)$, we have shown above that $-a$ is the largest eigenvlaue of *A* less than *0*. More generally, suppose that $A: R(A) \subset H \to H$ is a normal linear map with closed range. If H_c is the complexification of the real Hilbert space H and $A_c : D(A_c) \subset H_c \to H_c$ is defined by $A_c(x+iy) = Ax + iAy$ *on* $D(A_c) = \{x+iy | x, y \in D(A)\}$, then Hetzer [He] has shown that

$$a = a(A) = inf\{|\lambda|^2/(-Re\lambda) | \lambda \in \sigma(A_c), Re\lambda < 0\}.$$

In general a could be infinite, in which case A is a linear monotone map, and could belong to $\sigma(A_c)$ or be a regular value of A_c.

Theorem 4.4 has been proved by Brezis–Nirenberg when $N : H \to H$ is a monotone map and the partial inverse A^{-1} is compact using the Leray–Shauder and monotone operator theories. In view of the various examples of pseudo A–proper maps A+ N discussed in Section 2.2, this result holds also for many other classes of maps A and N, even when A^{-1} is not compact. For example, we have the following corollary.

COROLLARY 4.3 Let A and N satisfy conditions (4.13) - (4.14) and $N = N_1 + N_2$ be such that N_1 is c–strongly monotone, k_1–ball contractive and N_2 is k_2–ball contractive with k_1, k_2 sufficiently small and continuous. Then $A + N$ is surjective.

Proof. The map $A + N : D(A) \subset H \to H$ is A–proper by Proposition 2.6. Moreover, since N is c–strongly monotone, it is well known to be surjective and the conclusion follows from Theorem 4.4. ∎

For our second corollary, we need to recall the following result proven in [Mi–7].

PROPOSITION 4.1 Let $A : D(A) \subset H \to H$ be selfadjoint, $H^{\pm}$ be closed subspaces of H with $H = H^- \oplus H^-$ and $H^- \cap H^+ = \{0\}$ and $\Gamma = \{H_n, P_n\}$ be a scheme for H that satisfies (2.35). Suppose that $N : H \to H$ has a symmetric weak Gateaux derivative $N'(x)$ *on* H and there are symmetric maps $B^{\pm} \in L(H)$ such that $B^- \leq N'(x) \leq B^+$ for each $x \in H$ and

(4.15) $((A - B^-)x, x) \leq 0$ for $x \in D(A) \cap H^-$;

(4.16) $((A - B^+)x, x) \geq 0$ for $x \in D(A) \cap H^+$.

Then $\pm A + N : D(A) \subset H \to H$ is pseudo A–proper *w.r.t.*Γ.

In view of Proposition 4.1, we have the following special case of Theorem 4.4.

COROLLARY 4.4 Let $A : D(A) \subset H \to H$ and $N : H \to H$ be as in Proposition 4.1 and satisfy conditions (4.15) – (4.16). Then $Int(R(A) + convR(A)) \subset R(\pm A + N)$ and $\pm A + N$ is surjective if such is N.

Next, we shall discuss some conditions that imply (4.15) – (4.16). Let $A : D(A) \subset H \to H$ be selfadjoint and suppose that

(4.17) $B^{\pm} = \sum_{i=1}^{m} \lambda_i^{\pm} P_i^{\pm}$ commute with A, where $P_i^{\pm} : H \to \ker(B^{\pm} - \lambda_i^{\pm} I)$

are orthogonal projections, $\lambda_1^{\pm} \leq \ldots \leq \lambda_m^{\pm}$ and $\lambda_i^{\pm}$ are pairwise disjoint;

(4.18) $\bigcup_{i=1}^{m} [\lambda_i^-, \lambda_i^+] \subset \overline{\varrho(A)}$, the closure of the resolvent set of A.

As we have seen in Section 2.2, $A - B^{\pm}$ have the spectral resolution

$$A - B^{\pm} = \sum_{i=1}^{m} \int_{-\infty}^{\infty} (\lambda - \lambda_i^{\pm}) dE_\lambda P_i^{\pm}.$$

Let $\delta > 0$ be small enough such that $\mu_i^{\pm} = \lambda_i^{\pm} \mp \delta$ satisfy

$$\bigcup_{i=1}^{m} [\mu_i^-, \mu_i^+] \subset \varrho(A). \tag{4.19}$$

Then the operators $B_\delta^{\pm} = B^{\pm} \mp \delta I$ have $\mu_i^{\pm}$ as their eigenvalues and $\ker(B_\delta^{\pm} - \mu_i^{\pm}) = \ker(B^{\pm} - \lambda_i^{\pm})$. Since $B_\delta^{\pm}$ commute with A, the spectral resolutions of $A - B_\delta^{\pm}$ are

$$A - B_\delta^{\pm} = \sum_{i=1}^{m} \int_{-\infty}^{\infty} (\lambda - \mu_i^{\pm}) dE_\lambda P_i^{\pm}. \tag{4.20}$$

Define the orthogonal projections $P^{\pm}$ by

$$P^+ = \sum_{i=1}^{m} E(\mu_i^+, \infty) P_i^+ \quad \text{and} \quad P^- = \sum_{i=1}^{m} E(-\infty, \mu_i^-) P_i^-$$

and let $H^{\pm} = P^{\pm}(H)$. Note that by (4.19)

$$P^+ = \sum_{i=1}^{m} E(\mu_i^-, \infty)P_i^+ \tag{4.21}$$

and

$$dist(\bigcup_{i=1}^{m} [\mu_i^-, \mu_i^+], \sigma(A)) \geq \delta.$$

Moreover, it follows from (4.20) that $((A - B_\delta^-)x, x) \leq \delta \| x \|^2$ on $D(A) \cap H^-$ and $((A - B_\delta^-)x, x) \geq \delta \| x \|^2$ on $D(A) \cap H^+$. Hence, conditions (4.15) - (4.16) hold. Moreover, if we assume that $P_i^- = P_i^+$ for $1 \leq i \leq m$, then by (4.21), $P^+ = I - P^-$, *i.e.*, $H^+ = (H^-)^\perp$.

By the discussion above, we have

COROLLARY 4.5 Let $A : D(A) \subset H \to H$ be selfadjoint, $H^\pm$ be closed subspaces of H with $H = H^- \oplus H^+$ and $H^- \cap H^+ = \{0\}$ and $\Gamma = \{H_n, P_n\}$ satisfy (2.35). Suppose that $N : H \to H$ has a symmetric weak Gateaux derivative $N'(x)$ on H and $B^- \leq N'(x) \leq B^+$ for $x \in H$ and some selfadjoint maps $B^\pm \in L(H)$. If conditions (4.13) - (4.14) and (4.17) - (4.18) hold, then $Int(R(A) + convR(A) \subset R(\pm A + N)$ and $\pm A + N$ is surjective if such is N.

A particular case of Corollary 4.3 is when $B^- = \lambda_k I$ and $B^+ = \lambda_{k+1} I$ for two consecutive eigenvalues $\lambda_k < \lambda_{k+1}$ of A. If $H^-(resp. H^+)$ is the subspace of H spanned by the eigenvectors of A corresponding to the eigenvalues $\lambda_i \leq \lambda_k$ $(resp. \lambda_i \geq \lambda_{k+1})$, then $H^+ = (H^-)^\perp$ and $H = H^- \oplus H^+$. Let $\Gamma = \{H_n, P_n\}$ be a scheme for H with $P_n Ax = Ax$ and $P^\pm : H \to H^\pm$ be the orthogonal projections onto $H^\pm$. Then Γ satisfies (2.35) with $H_n = H_n^- \oplus H_n^-$ and $H_n^\pm = H_n \cap H^\pm$.

Now, since $B^- \leq N'(x) \leq B^+$ on H, the mean value theorem implies that $\| Nx - Ny \| \leq \max \{ \| B^- \|, \| B^+ \| \} \| x - y \|$ for $x, y \in H$. In particular, taking $B^- = \lambda_k I$ and $B^+ = \lambda_{k+1} I$, we get

$$\limsup_{\| x \| \to \infty} \| Nx \| / \| x \| \leq \max \{ |\lambda_k|, |\lambda_{k+1}| \},$$

and, by (4.14) as noted above

$$\limsup_{\| x \| \to \infty} \| Nx \| / \| x \| \leq \gamma < a \ ,$$

Hence, for N to interact with λ_k(*or* λ_{k+1}), it must be the zero eigenvalue.

4.2 A Perturbation Method

In this section we shall present another way of studying Eq. (1.1) by looking at the perturbed equations

$$Ax + Nx + \epsilon Gx = f, \quad \epsilon > 0 \tag{4.22}$$

where G is a bounded map. This approach consists of three basic steps. The first step is to establish the solvability of (4.22) for each $\epsilon \in (0, \epsilon_0)$; the second step is to obtain a priori bound on the solutions, i.e. to show that the set $\{ x_\epsilon | x_\epsilon$ is a solution of (4.22)$\}$ is bounded in a suitable space as $\epsilon \to 0$; and the final step is to show that a weak limit of $\{ x_\epsilon \}$ is a solution of $Ax + N x = f$.

The approach requires a closedness property of $A + N$, which is defined next.

Definition 4.1 Let X be a Banach space embedded in H. We say that $A + N : D(A) \cap X \to H$ satisfies condition (*) if whenever $\{x_\epsilon | Ax_\epsilon + Nx_\epsilon + \epsilon Gx_\epsilon = f, 0 < \epsilon < \epsilon_0\}$ is bounded as $\epsilon \to 0$ (in X or H), then there is an $x \in D(A)$ such that $Ax + Nx = f$. (Usually, x is a weak limit of $\{x_\epsilon \}$.)

The following result gives various conditions on A and N which guarantee that condition (*) holds.

PROPOSITION 4.2 Let $A : D(A) \cap X \to H$ be a closed linear densely defined map and N be a nonlinear map. Let the conditions of either one of Propositions 2.1 - 2.6 or Proposition 4.1 hold. Then $A + N$ satisfies condition (*).

Proof. (Sketch) Let $\{x_\epsilon | Ax_\epsilon + Nx_\epsilon + \epsilon Gx_\epsilon = f, 0 < \epsilon < \epsilon_0\}$ be bounded in a corresponding space. Then, using similar arguments as in the proofs Propositions 2.1 - 2.6 or Proposition 4.1, one can show that the weak limit x of $\{x_\epsilon \}$ solves the equation $Ax + Nx = f$. ■

The problem of getting a priori estimates for (4.22), i.e. of showing the boundedness of $\{x_\epsilon \}$, is harder to handle. Our basic result is [Mi–6]:

THEOREM 4.5 Let $A: D(A) \subset H \to H$ be a linear densely defined selfadjoint map with $R(A) = N(A)^{\perp}$ and $a > 0$ be the largest number such that $(Ax, x) \geq -a^{-1} \| Ax \|^2$ on $D(A)$. Suppose that $N: H \to H$ is nonlinear, condition (*) holds and

(4.23) There is a decomposition $f = f^* + f^{**}$ with $f^{**} \in R(A)$ such that for some $\gamma < a$ and a constant c

$$(Nx - f^{**}, x) \geq \gamma^{-1} \| Nx \|^2 - c \quad \text{for} \quad x \in H,$$

(4.24) Eq. (4.22) with $G = I$ is solvable for each $0 < \epsilon < \epsilon_0$, and either

(4.25) the set $\{x_\epsilon | Ax_\epsilon + Nx_\epsilon + \epsilon x_\epsilon = f, \| Ax_\epsilon \| \leq C, \| Nx_\epsilon \| \leq C$ for all $0 < \epsilon < \epsilon_0$ and some $C\}$ is bounded in H, or

(4.26) $\| N(x_0 + x_1) \| \to \infty$ *as* $\| x_0 \| \to \infty, x_0 \in N(A)$, uniformly for x_1 in bounded subsets of $R(A)$.

Then Eq. (1.1) is solvable.

Proof. Let $\epsilon > 0$ be small and $Ax_\epsilon + Nx_\epsilon + \epsilon x_\epsilon = f$ for some $x_\epsilon \in D(A)$. We need to show that $\{x_\epsilon\}$ is bounded as $\epsilon \to 0$. Taking the inner product with x_ϵ, we find

$$\epsilon \| x_\epsilon \|^2 + (Nx_\epsilon - f, x_\epsilon) = -(Ax_\epsilon, x_\epsilon) \leq a^{-1} \| Ax_\epsilon \|^2,$$

and by (4.23),

$$\epsilon \| x_\epsilon \|^2 + \gamma^{-1} \| Nx_\epsilon \|^2 \leq -(Ax_\epsilon, x_\epsilon) + (f^{**}, x_\epsilon) + C \| Nx_\epsilon \| + C.$$

Since $Av = f^*$ for some $v \in D(A)$, and $(A(x_\epsilon - v), x_\epsilon - v) \geq -a^{-1} \| A(x_\epsilon - v) \|^2$, we get

$$\begin{aligned} \epsilon \| x_\epsilon \|^2 + \gamma^{-1} \| Nx_\epsilon \|^2 &\leq a^{-1} \| A(x_\epsilon - v) \|^2 - (A(x_\epsilon - v), v) + C \| Nx_\epsilon \| + C \\ &\leq a^{-1} \| Ax_\epsilon \|^2 + C \| Ax_\epsilon \| + C \| Nx_\epsilon \| + C. \end{aligned}$$

For each $\gamma < \beta < a$ there is a $C(\beta)$ such that

$$\gamma^{-1} \| Nx_\epsilon \|^2 - C \| Nx_\epsilon \| \geq \beta^{-1} \| Nx_\epsilon \|^2 - C(\beta).$$

Similarly, for each $\beta < \delta < a$ there is a $C(\delta)$ such that

$$\begin{aligned} \epsilon \| x_\epsilon \|^2 + \beta^{-1} \| Nx_\epsilon \|^2 &\leq a^{-1} \| Ax_\epsilon \|^2 + C \| Ax_\epsilon \| + C(\beta) \\ &\leq \delta^{-1} \| Ax_\epsilon \|^2 + C(\beta, \delta), \end{aligned}$$

where $C(\beta, \delta) = C(\beta) + C(\delta)$. Since $Nx_\epsilon = f - Ax_\epsilon - \epsilon x_\epsilon$, the last inequality implies that

$$\begin{aligned} &\epsilon[(1 + \epsilon\beta^{-1}) \| x_\epsilon \|^2 - 2\beta^{-1} \| f \| \| x_\epsilon \|] + \\ &(\beta^{-1} - \delta^{-1} - 2\epsilon(a\delta)^{-1}) \| Ax_\epsilon \|^2 - 2\delta^{-1} \| f \| \| Ax_\epsilon \| \leq C(\beta, \delta). \end{aligned} \tag{4.27}$$

This implies that for some sufficiently small constants $C_i > 0, i = 1, 2, 3,$ independently of ϵ,

$$\epsilon C_1 \| x_\epsilon \|^2 + C_2 \| Ax_\epsilon \|^2 \leq C_3,$$

and therefore $\| Ax_\epsilon \| \leq C$ for all small ϵ. Moreover, $\| Nx_\epsilon \| \leq C$ by (4.27) and consequently $\| x_\epsilon \| \leq C$ for all small ϵ if $A + N$ satisfies (4.25). Thus, $Ax_\epsilon + Nx_\epsilon = f - \epsilon x_\epsilon \to f$ as $\epsilon \to 0$ and $Ax + Nx = f$ for some $x \in D(A)$ by condition (*).

Next, if instead of (4.25) we assume (4.26), then $x_\epsilon = x_{0\epsilon} + x_{1\epsilon}$ and $\| x_{1\epsilon} \| = \| A^{-1}Ax_\epsilon \| \leq C \| A^{-1} \|$ for all small ϵ. Moreover, $\| x_{0\epsilon} \| \leq C$ for such ϵ by (4.26), and consequently $\{x_e\}$ is bounded. Then, the conclusion follows as above. ∎

When N is a gradient map, we have the following sufficient and necessary condition for the solvability of (1.1).

THEOREM 4.6 Let $A : D(A) \subset H \to H$ be a linear selfadjoint map and $a > 0$ be the largest number such that $(Ax, x) \geq -a^{-1} \| Ax \|^2$ on $D(A)$. Assume that either (4.25) or (4.26) holds, $N = \partial F$ for some convex function $F : H \to R$ and

(4.28) $\lim sup \| Nx \| / \| x \| < a/2, \; as \; \| x \| \to \infty.$

Suppose that either one of the following conditions holds

(4.29) $A^{-1} : R(A) \subset H \to H$ is compact,

(4.30) $0 \in \sigma(A)$ and $\sigma(A) \cap (0, \infty) \neq \phi$ and consists of eigenvalues of finite multiciplicities,

(4.31) There are closed subspaces $H^{\pm}$ of H with $H = H^- \oplus H^+$ and $H^- \cap H^+ = \{0\}$, a scheme $\Gamma = \{H_n, P_n\}$ satisfying (2.35) and N has a symmetric weak Gateaux derivative $N'(x)$ on H such that for some symmetric maps $B^{\pm} \in L(H)$ with $B^- \leq N'(x) \leq B^+$ on H we have

(i) $((A - B^-)x, x) \leq 0$ for $x \in D(A) \cap H^-$;

(ii) $((A - B^+)x, x) \geq 0$ for $x \in D(A) \cap H^+$.

Then, Eq. (1.1) is solvable if and only if $f = f^* + f^{**}$ with $f^* \in R(A)$ and $f^{**} \in R(N)$.

Proof. If x is a solution of $Ax + Nx = f$, then $f^* = Ax$ and $f^{**} = Nx$. Conversely, let $f = f^* + f^{**}$ with $f^* \in R(A)$ and $f^{**} \in R(N)$. Then (4.28) and Proposition A.4 in [Br–Ni–1] imply that N and $N_\epsilon = N + \epsilon I$ satisfy (4.14) and (4.23), respectively, for each $\epsilon > 0$ small. Moreover, N_ϵ is monotone and $\| N_\epsilon x \| \to \infty$ as $\| x \| \to \infty$, and therefore it is surjective.

Since $A + N_\epsilon$ is pseudo A–proper by Proposition 2.2 and 4.1, Eq. (1.1) is solvable for each small $\epsilon > 0$ by Theorem 4.4. Finally, since condition (*) holds by Proposition 4.2, Eq. (1.1) is solvable by Theorem 4.5. ■

When $H = L_2(Q, R^m)$ we can relax condition (4.26) in Theorem 4.5 (cf. [Mi–7]).

COROLLARY 4.6 Let $Q \subset R^n$ be a bounded domain, $H = L_2(Q, R^m)$ and $Nu = D_u F(x, u)$ with $F : QxR^m \to R$ measurable in x and convex and C^1 in u. Then, condition (4.26) in Theorem 4.5 can be weakened to

$$(4.32) \qquad \int_Q F(x, u_0(x))dx \to \infty \quad \text{as} \quad \| u_0 \|_{L_2} \to \infty, u_0 \in N(A).$$

Remark 4.2 When $n = 1$, $a/2$ in (4.28) can be replaced by a by Proposition A.6 in [Br–Ni–1].

V. PERIODIC SOLUTIONS OF SEMILINEAR WAVE EQUATIONS AT RESONANCE

In this section we shall apply some of the abstract results from the previous section to semilinear (systems of) wave equations (3.6) allowing some type of interaction of the nonlinearity F with the spectrum of the linear problem.

We begin with strongly monotone and contractive nonlinearities. Suppose $\Omega = (0, T)xQ$ and

(5.1) $F_i : \Omega x R^m \to R^m$ be Caratheory functions, $i = 1, 2$, such that for some positive constants k_i and c, with k_1, k_2 sufficiently small,

$$(F_1(t,x,y) - F_1(t,x,z)) \cdot (y - z) \geq c|y - z|^2$$

and

$$|F_i(t,x,y) - F_i(t,x,z)) \leq k_i|y - z|$$

for $a.e.$ $(t,x) \in \Omega$ and all $y, z \in R^m$.

(5.2) There are $a > 0$ and $h \in L_2(\Omega)$ such that for $\gamma_1 = 1$, and $\gamma_2 \in (0,1)$

$$|F_i(t,x,y)| \leq a|y|^{\gamma_i} + h(t,x) \quad \text{for} \quad a.e. (t,x) \in \Omega, y \in R^m.$$

Let L and V be as introduced in Section 3.1. Then, we have shown there that a linear map $A : D(A) \subset L_2 \to L_2$, induced by (3.6), is selfadjoint and $R(A) = N(A)^{\perp}$. Moreover, condition (4.13) holds for some $a_{\pm}$ and a_0.

We have

THEOREM 5.1 Let (5.1) – (5.2) hold with $a < a_+/2$ and $m > 1$ and $F = \partial\psi$ for some function $\psi(t,x,y) : \Omega x R^m \to R$ measurable in (t,x) and differentiable and convex in y. Then there is a T-periodic weak solution $u \in L_2$ for each $f \in L_2$.

Proof. Define $N_i u = F_i(t,x,u)$ for $u \in L_2$ and let $N = N_1 + N_2$. By (5.1), N_1 is c-strongly monotone and k_1-ball contractive and N_2 is k_2-ball-contractive.

Hence, $A + N$ is A-proper $w.r.t.$ $\Gamma = \{H_n, P_n\}$ for L_2 with $P_n Au = Au$ for $u \in H_n$ by Proposition 2.6. Moreover, $\| N_2 u \| / \| u \| \to 0$ as $\| u \| \to \infty$ and

$$(Nu, u) = (N_1 u, u) - \| N_2 u \| \| u \| \geq c \| u \|^2 - (\| N_1 0 \| + \| N_2 u \|) \| u \| .$$

Thus, N is coercive, i.e. $(Nu, u)/ \| u \| \to \infty$ as $\| u \| \to \infty$, and monotone and therefore, it is surjective.

Next, by Proposition A.4 in [Br–Ni–1], there is a $\gamma < a_+$ such that

$$(Nu - Nv, u) \geq \gamma^{-1} \| Nu \|^2 - c(v) \quad \text{for all} \quad u, v \in L_2$$

and some constant $c(v)$. Hence, by Theorem 4.4, the equation $Au - Nu = f$ is solvable for each $f \in L_2$, i.e., there is a T-periodic weak solution $u \in L_2$ of (3.6) for each $f \in L_2$. ∎

As will be seen below, conditions on F can be greatly relaxed when $m = 1$. Next, we shall study the solvability of (3.6) for a given f.

THEOREM 5.3 Let $G(t, x, u) : \Omega x R^m \to R$ be T-periodic in t, measurable in (t,x) and convex and C^1 in u and $F(t, x, u) = D_u G(t, x, u)$ satisfy (5.2). Suppose that

$$\int_\Omega G(t, x, u) dt dx \to \infty \quad \text{as} \quad \| u_0 \|_{L_2} \to \infty, u_0 \in N(A) \tag{5.3}$$

and either $a < a_+$ if $m = 1$, or $a < a_+/2$ and (3.20) – (3.21) and (4.18) hold and b^- and b^+ commute if $m > 1$. Then (3.6) has a T-periodic weak solution $u \in L_2$ if and only if $f = f^* + f^{**}$ with $f^* \in R(A)$ and $f^{**} \in R(N)$, where $Nu = F(t, x, u)$.

Proof. Let $m = 1$ and $a < a_+$. Then $A^{-1} : R(A) \subset L_2 \to L_2$ is compact and N satisfies condition (4.14) by Proposition A.1 in [Br–Ni–1]. If $m > 1$ and $a < a_+/2$, condition (4.14) holds by Proposition A.4 in [Br–Ni–1]. Hence, in either case, by our discussion in Section 4.2 the conclusion of the theorem follows from Theorem 4.6 and Corollary 4.6. ∎

COROLLARY 5.1 Let $n = 1$ and $F(t,x,u) = D_u G(t,x,u)$ satisfy (5.2) with $h \in L_\infty(\Omega), a < a_+$, and

(5.4) $\quad G(t,x,u) \to \infty$ as $|u| \to \infty$ uniformly a.e. in Ω.

Then there is 2π periodic weak solution $u \in L_2$ of

$$(5.5) \quad \begin{cases} u_{tt} + u_{xxxx} + F(t,x,u) = f(t,x), & t \in R, x \in (0,\pi), \\ u(t,0) = u(t,\pi) = 0, & t \in R \\ u_{xx}(t,0) = u_{xx}(t,\pi) = 0 & \\ u(0,x) - u(2\pi,x) = u_t(0,x) = u_t(2\pi,x) = 0, & t \in R, x \in (0,\pi) \end{cases}$$

if and only if $f = f^* + f^{**}$ with $f^* \in R(A)$ and $f^{**} \in R(N)$.

Proof. Since $h \in L_\infty$ and G is convex in u, as in [Ma–Wi] there is a $\delta > 0$ such that

$$G(t,x,u) \geq \delta|u| - h(t,x) \quad \text{a.e.}\,(t,x) \in \Omega, u \in R^m. \tag{5.6}$$

Since the L_1 and L_2 norms are equivalent on $N(A)$ (*cf.* [Ba–Sa]), this implies (5.3) and the conclusion follows from Theorem 5.3. ∎

When $f = 0$, we refer to [Ma–Wi] for related results obtained by a dual variational method. When $n = 1$, our results imply the following ones of Bahri–Brezis [Ba–Br] and Bahri–Sanchez [Ba–Sa] for the wave equation and (5.5), respectively.

COROLLARY 5.2 Let $g : R \to R$ be continuous nondecreasing and for some $a < 3$, b and $\delta > 0$

(5.7) $\quad |g(u)| \leq a|u| + b$ for all $u \in R$,

$$f = f^* + f^{**} \in L_\infty(\Omega) \text{ with } f^* \in R(A) \text{ and}$$
$$g(-\infty) + \delta \leq f^{**}(t,x) \leq g(+\infty) - \delta \text{ a.e. in } \Omega. \tag{5.8}$$

Then there is a weak solution of (5.5) with F replaced by g, and of

$$\begin{cases} u_{tt} - u_{xx} + g(u) = f(t,x), & t \in R, x \in (0,\pi) \\ u(t,0) = u(t,\pi) = 0 & t \in R \\ u(t+2\pi, x) = u(t,x), & t \in R, x \in (0,\pi). \end{cases} \tag{5.9}$$

Proof. Consider first (5.5). We note that it is equivalent to

$$Av + g(u^* + v) - f^{**} = Av + D_v G(t,x,v) = 0 \tag{5.10}$$

where $u^* \in L_\infty(\Omega)$ is the solution of $Au = f^*, u = u + v$, and

$$G(t,x,v) = \int_0^v [g(u^* + s) - f^{**}(t,x)]ds.$$

Since (5.7) implies

$$(g(u) - f^{**}(t,x))u \geq \delta|u|/2 - C \text{ for } (t,x) \in \Omega, u \in R,$$

we have that (cf. [Mi–7]) (5.6) and therefore, (5.3) holds. Hence, the weak solvability of (5.5) follows from Theorem 5.3 applied to (5.10).

Next, consider (5.9). Conditions (5.7) – (5.8) and the monotonicity of g imply (4.23), (4.24) and (*) and by Theorem 4.4 and Proposition 4.2. As shown by the proof in [Br], condition (4.25) also holds, and consequently the weak solvability of (5.9) follows from Theorem 4.5. ■

Another condition on f^{**} is given in the following result.

THEOREM 5.4 Let the conditions of Theorem 5.3 hold. Suppose that $f = f^* + f^{**} \in L_2(\Omega)$ with $f^* \in R(A)$ and

$$(F(t,x,y) - f^{**}(t,x)) \cdot y \geq c\left| F(t,x,y) - f^{**}(t,x)\right| |y| \tag{5.11}$$

for all $|y| \geq R$, a.e. $(t,x) \in \Omega$, some R, and a sufficiently small $c > 0$. Then there is a T-periodic weak solution of (3.6).

Proof. We shall first show that (5.2) and (5.11) imply condition (4.23). We have

$$(Nu - f^{**}, u) = \int_{|u(t,x)| < R} (F(t,x,u(t,x)) - f^{**}(t,x)) \cdot u(t,x)dtdx$$

$$+ \int_{|u(t,x)| \geq R} (F(t,x,u(t,x)) - f^{**}(t,x)) \cdot u(t,x)dtdx = I_1 + I_2.$$

By (5.2),

$$|I_1| \leq \int_{|u(t,x)| < R} (a|u(t,x)| + h(t,x) + |f^{**}(t,x)|)|u(t,x)|dtdx$$

$$\leq aR^2 + R(\| h \| + \| f^{**}\|) = c_1.$$

By (5.2) and (5.11), for $|u(t,x)| \geq R$

$$(F(t,x,u(t,x)) - f^{**}(t,x)) \cdot u(t,x) \geq \frac{c}{a}| F(t,x,u(t,x)) - f^{**}(t,x)|.$$

$$(|F(t,x,u(t,x))| - h(t,x))$$

$$= \frac{c}{a}|F(t,x,u(t,x))|^2 - \frac{c}{a}|F(t,x,u(t,x))|(|f^{**}(t,x)| - h(t,x))$$

$$- \frac{c}{a}|f^{**}(t,x)|h(t,x).$$

Hence,

$$I_2 \geq \frac{c}{a}\| Nu \|^2 - \left(\frac{c}{a}\| f^{**}\| + \| h \|\right)\| Nu \| - \frac{c}{a}\| f^{**}\| \quad \| h \| .$$

Since $C \leq \epsilon C^2 + c(\epsilon)$ for each $\epsilon > 0$, condition (4.23) follows easily from the estimates on I_1 and I_2.

Next, in view of Propositions A.1 and A.4 in [Br–Ni–1], condition (4.14) holds and therefore the equation $Ax + Nx + \epsilon x = f$ is solvable for each $\epsilon > 0$ as shown in the proof of Theorem 4.6. Hence by our discussion in Section 4.2, the conclusion follows from Theorem 4.5. ■

Remark 5.1 It is easy to see that condition (5.8) implies condition (5.11).

For our final result, we assume that $\Omega = (0, T)x(0, \pi)$, and $F : \Omega x R \to R$ is a Caratheodory function such that for some $h \in L_2(\Omega)$

$$|F(t, x, y)| \leq h(t, x) \text{ for a.e. } (t, x) \in \Omega, y \in R. \tag{5.12}$$

Let $F_{\pm}(t, x) = \lim F(t, x, y)$ as $y \to \pm \infty$ for $(t, x) \in \Omega$. We denote by $Au = u_{tt} - u_{xx}$ the operator acting on $u \in L_2$ that satisfy the Dirichlet boundary conditions in x. We have

THEOREM 5.5 [Mi–9] Let (5.12) hold and for some $f_0 \in \ker A$ and all $u \in \partial B(0, 1) \cap \ker A$, either

$$(5.13) \qquad \int_{u>0} F_+(t, x)u(t, x)dtdx + \int_{u<0} F_-(t, x)u(t, x)dtdx > \int_{\Omega} f_0(t, x)dtdx,$$

or

$$(5.14) \qquad \int_{u>0} F_+(t, x)u(t, x)dtdx + \int_{u<0} F_-(t, x)u(t, x)dtdx < \int_{\Omega} f_0(t, x)dtdx.$$

Then there is a T–periodic weak solution of (5.9) (with g replaced by F) for each $f \in f_0 \oplus R(A)$.

The proof of this result is based on Theorem 4.3 and can be found in [Mi–9]. It improves Corollary 2.8 in Brezis–Nirenberg [Br–Ni], where it is shown that $f \in \overline{R(A + N)}$ provided $F(t, x, y)$ is monotone and has a linear growth in y and with the equality sign also allowed in (5.13) – (5.14).

REFERENCES

[A] S. Ahmad (1973). An existence theorem for periodically perturbed conservative systems, Mich. Math., I., 10, 385–392.

[Am–1] H. Amann (1979). Saddle points and multiple solutions of differential equations, Math. Zeitschr.,169, 127–166.

[Am–2] H. Amann (1982). On the unique solvability of semilinear operator equations in hilbert spaces, J. Math. Pures et Appl., 61, 149–175.

[A–Z] H. Amann and E. Zehnder (1980). Nontrivial solutions for a class of nonresonance problems and applications to nonlinear differential equations, Annali Scuola Norm. Sup. Pisa, Ser. IV, 7, 539–603.

[Ba–Br] A. Bahri and H. Brezis (1980). Periodic solutions of a nonlinear wave equation, Proc. Royal Soc. Edinburg, A85, 313b–320.

[Ba–Sa] A. Bahri and S. Sanchez (1981). Periodic solutions of a nonlinear telegraph equation in one dimension, Bull. Un. Met. Ital., t.5, 18–B, 709–720.

[Be–Fo] Y. Benci and D. Fortunato (1982). The dual method in critical point theory–multiplicity results for indefinite functionals, Ann. Mat. Pura Appl., 32, 215–242.

[Ber–Fi] H. Beresticky and D. de Figueiredo (1981). Double resonance in semilinear elliptic problem, Comm. Diff. Eq., 6, 91–120.

[Br] H. Brezisd (1983). Periodic solutions of nonlinear vibrating strings and duality principles, Bull. Amer. Math. Soc., 8, 409–426.

[Br–Ni1] H. Brezis And L. Nirenberg (1978). Characterizations of the ranges of some nonlinear operators and applications to boundary value problems, Ann. Scuola Norm. Sup. Pisa, 5, 225–326.

[Br–Ni2] H. Brezis and L. Nirenberg (1978). Forced vibrations for a nonlinear wave equation, Comm. Pure Appl. Math., 31, 1–30.

[Bro–Li] K.J. Brown and S.S. Lin (1980). Periodically perturbed conservative systems and a global inverse function theorem, Nonlinear Analysis, TMA, 4, 193–201.

[Ce–Ka] L. Cesari and R. Kannan (1982). Solutions of nonlinear hyperbolic equations at resonance, Nonlinear Analysis, TMA, 6, 751–805.

[Ch–Ho] K.C. Chang and C.W. Hong (1985). Periodic solutions for the semilinear spherical wave equation, Acta Math. Sinica, 1, 87–96.

[Co] J.M. Coron (1983). Periodic solutions of nonlinear wave equations without assumption of monotonicity, Math. Ann., 262 273–285.

[Fig] D.de Figueiredo (1974). The range of nonlinear operators with linear asymptotes which are not invertible, Comm. Math. Univ. Carolinae, 15, 415–428.

[Fi] P.M. Fitzpatrick (1978). Existence results for equations involving noncompact perturbations of Fredholm mappings with applications to differential equations, J. Math. Anal. Appl., 66, 151–177.

[He] G. Hetzer (1984). A spectral characterization of monotonicity properties of normal linear operators with an application to nonlinear telegraph equations, J. Operator Theory, 12, 333–341.

[Ho] H. Hofer, (1982). On the range of a wave operator with non-monotone nonlinearity, Math. Nachr., 106, 327–340.

[Kr–Za] M.A. Krasnoselskii and P.O. Zabreiko (1984). Geometrical Methods of Nonlinear Analysis, Springer–Verlag.

[La–La] E. Landesman and A. Lazer (1970). Nonlinear perturbations of linear elliptic boundary value problems at resonance, J. Math. Mech., 19, 609–623.

[La] A.C. Lazer (1972). Applications of a lemma on bilinear forms to a problem in nonlinear oscillations, Proc. Amer. Math. Soc., 33, 89–94.

[Ma–1] J. Mawhin (1977). Periodic solutions on nonlinear telegraph equations, Dynamical Systems, Bednarek–Cesari Ed., Acad. Press, NY, 193–210.

[Ma–2] J. Mawhin (1981). Compacticite, Monotonic et Convexite dans L'etude de Problemes aux Limites Semi–lineaires, Lecture Notes, Univ. de Sherbrooke.

[Ma–3] J. Mawhin (1981). Conservative systems of semi–linear wave equations with periodic–direchlet boundary conditions, J. Diff. Equations, 42, 116–128.

[Ma–4] J. Mawhin (1982). Nonlinear functional analysis and periodic solutions of semilinear wave equations, in Nonlinear Phenomena In Math. Sci., Arlington, Lakshmikantham ed., Acad. Press, NY, 671–681.

[Ma–Wa] J. Mawhin and J. Ward (1983). Asymptotic nonuniform nonresonance conditions in the periodic - dirichlet problem for semilinear wave equations, Annali Mat. Pura Appl. 135, 85–97.

[Ma–Wi] J. Mawhin and M. Wilhem (1986). Convex perturbations of quadratic forms, Ann. de L'institut Henri Poincare–Analyse nonlineaire, 3 (6) 431–453.

[Mi–1] P.S. Milojević (1980). Continuation theorems and solvability of equations involving nonlinear noncompact perturbations of Fredholm mappings, Proc. 12th Seminario Brasileiro de Analize, ITA, Sao Jose dos Campos, 163–189.

[Mi-2] P.S. Milojević (1983). Approximation - solvability results for equations involving nonlinear perturbations of Fredholm mappings with applications to differential equations, (G.Zapata ed.) Proc. Int. Sem. Funct. Anal. Holom. Approx. Theory (Rio de Janeiro, August 1979), Lecture Notes in Pure and Appl. Math., Vol. 83, M. Dekker, NY, 305-358.

[Mi-3] P.S. Milojević (1982). Continuation theory for a-proper and strongly a-closed mappings and their uniform limits and nonlinear perturbations of Fredholm mappings, (E. Barroso ed.), Proc. Int. Sem. Funct. Anal. Holom. Approx. Theory. (Rio De Janeiro, August 1980), Mathematics Studies, Vol. 71, North-Holland, Amsterdam, 299-372.

[Mi-4] P.S. Milojević (1984). Approximation-solvability of some noncoercive nonlinear equations and semilinear problems at resonance with applications, (G. Zapata, ed.) Proc. Int. Sem. Funct. Anal. Holom. Approx. Theory (Rio de Janiero, August 1981), Mathematics Studies, Vol. 86, North-Holland, Amsterdam, 259-295.

[Mi-5] P.S. Milojević (1986). On the index and the covering dimension of the solution set of semilinear equations, Proc. Symp. Pure Math., 45 (2), Amer. Math. Soc., 183-205.

[Mi-6] P.S. Milojević (1988). Solvability of nonlinear operator equations with applications to hyperbolic equations, (B. Stankovic et. al., eds.) Proc. Generalized Funct. Convergence Structures and their Applic., Plenum Press, NY, 245-250.

[Mi-7] P.S. Milojević, Solvability of semilinear hyperbolic equations at resonance, Proc. Diff. Eq. and Applic. Ohio University, March 1988 (to appear).

[Mi-8] P.S. Milojević, Periodic solutions of semilinear hyperbolic equations at resonance, J. Math. Anal. Appl. (to appear).

[Mi-9] P.S. Milojević, Nonresonance periodic-boundary value problems for semilinear hyperbolic equations (to appear).

[Mi-10] P.S. Milojević, Period solutions of systems of semilinear hyperbolic equations (to appear).

[Mi-11] P.S. Milojević, Error estimates for the a-proper, mapping method involving non-differentiable operators with applications to nonlinear partial differential equations (to appear).

[P-S] D. Pascali and S. Sburlan (1978). Nonlinear Mappings of Monotone Type, Ed. Acad., RSR & Sijthoff - Noordhoff Int. Publ.

[Pet] W.V. Petryshyn (1975). On the approximation - solvability of equations involving a-proper and pseudo a-proper mappings, Bull. Amer. Math. Soc., 81, 221-312.

[Per] A.I. Perov (1953). On the principle of the fixed point with two-sided estimates, Dokl. Acad. Nauk SSSR, 124, 756-759.

[Pro] G. Prodi (1956). Soluzioni periodiche di equazioni a derivate parziali di tipo iperbolico non lineari, Ann. di Mat. Pura Applic., 42, 25-49.

[Ra] P. Rabinowitz (1967). Periodic solutions of nonlinear hyperbolic partial differential equations, Comm. Pure Appl. Math., 20, 145-205.

[Sm] M.W. Smiley (1987). Eigenfunction methods and nonlinear hyperbolic boundary value problems at resonance, J. Math. Anal. Appl., 122, 129-151.

[Ta-1] K. Tanaka (1985). On the range of wave operators, Tokyo, J. Math., 8(2), 377-387.

[Ta-2] K. Tanaka (1985). Infinitely many periodic solutions for the equation: $u_{tt} - u_{xx} \pm |u|^{p-1}u = f(x,t)$, Comm. Partial Differential Equations, 10, 1317-1345, Part II, Trans. Amer. Math Soc. (to appear).

[Tar] E. Tarafdar (1983). An approach to nonlinear elliptic boundary value problems, J. Austral. Math. Soc., 34, 316-335.

[V] O. Vejvoda et al. (1981). Partial Differential Equations: Time-periodic Solutions, Noordhoff, Groningen.

[W] M. Wilhem (1981). Density of the range of potential operators, Proc. Amer. Math. Soc., 83, 341–344.

[Z] Z. Zhou (1987). The existence of periodic solutions of nonlinear wave equations on S^n, Comm. Part. Diff. Eq., 12(8), 829–882.

Some Remarks on Operator-Valued Means

ROGER D. NUSSBAUM Mathematics Department, Rutgers University, New Brunswick, New Jersey

1. INTRODUCTION

There is an enormous classical literature concerning what might be called "means and their iterates." The survey article [7] by Arazy, Claesson, Janson and Peetre provides an excellent bibliography. In this section we shall review some of these classical examples, particularly the arithmetic–geometric mean. The main body of the paper comprises results and open questions related to the problem of extending the classical means, particularly the arithmetic–geometric mean, to the case that the variables are (noncommuting), positive semi–definite, self–adjoint bounded linear operators on a Hilbert space.

We begin with the definition of the arithmetic–geometric mean. If a and b are positive real numbers, define $f(a,b) = (a_1,b_1)$ by

$$f(a,b) = (\frac{a+b}{2}, \sqrt{ab}) \tag{1.1}$$

If f^k denotes the kth iterate of f, it is a relatively easy calculus exercise to prove that there exists a positive number λ, dependent on a and b, such that

$$\lim_{k\to\infty} f^k(a,b) = (\lambda,\lambda) \tag{1.2}$$

Partially supported by NSF DMS 88 05395

With a little more work one can prove that the rate of convergence in eq. (1.2) is extremely rapid: technically, one has "quadratic convergence" in the sense of [31, Chapter 12]. The number λ in eq. (1.2) is called the arithmetic–geometric mean of a and b or AGM of a and b.

A much deeper fact about the AGM was discovered independently first by Landen, then by Lagrange and finally by Gauss. For a and b positive numbers, define $I(a,b)$ by

$$I(a,b) = \int_0^{\frac{\pi}{2}} (a^2\cos^2\theta + b^2\sin^2\theta)^{-\frac{1}{2}}\, d\theta.$$

Then, if f is as in eq. (1) one has that

$$I(a,b) = I(f(a,b)). \tag{1.3}$$

It follows from eq. (1.3) and eq. (1.2) that

$$I(a,b) = \lim_{k\to\infty} I(f^k(a,b)) = I(\lambda,\lambda) = \frac{\pi}{2\lambda}, \tag{1.4}$$

which provides an explicit formula for λ. More importantly, because λ can easily be computed to high accuracy with the aid of eq. (1.2), eq. (1.4) provides an efficient means of computing $I(a,b)$.

Further historical references and a variety of deeper results about the AGM can be found in [15].

There are many other examples of "iterated means", although explicit formulas as in eq. (1.4) are known in very few cases: see [13]. Thus, suppose that α and β are fixed positive reals such that $0 < \alpha, \beta < 1$. For a and b positive reals, define a map $f(a,b)$ by

$$g(a,b) = (\alpha a + (1-\alpha)b,\ a^\beta b^{1-\beta}) \tag{1.5}$$

Then one can prove that there exists a positive number μ (dependent on a and b) such that

$$\lim_{k\to\infty} g^k(a,b) = (\mu,\mu). \tag{1.6}$$

Convergence to μ in eq. (1.6) is at a quadratic rate if $\alpha = \beta$.

As a third example, define for $p \in \mathbb{R}$ and a and b positive reals,

$$M_p(a,b) = (\tfrac{1}{2}\, a^p + \tfrac{1}{2} b^p)^{\frac{1}{p}}. \tag{1.7}$$

If $p = 0$, one defines

$$M_0(a,b) = \lim_{p\to 0} M_p(a,b) = \sqrt{ab}.$$

For fixed real numbers p and q, define a map h by

$$h(a,b) = (M_p(a,b),\ M_q(a,b)).$$

Again, one can prove that for positive numbers a and b, there exists a positive number ν (dependent on a and b) such that

$$\lim_{k\to\infty} h^k(a,b) = (\nu,\nu).$$

There are examples of iterated means which involve functions of more than two variables. For example Borchardt [7,11] has considered the following map of four positive variables.

$$\begin{aligned} h(a,b,c,d) &= (a_1,b_1,c_1,d_1) \\ &= (\frac{a+b+c+d}{4}, \frac{\sqrt{ab} + \sqrt{cd}}{2}, \frac{\sqrt{ac} + \sqrt{bd}}{2}, \frac{\sqrt{ad} + \sqrt{bc}}{2}). \end{aligned}$$

Notice that if $a = c$ and $b = d$, then $a_1 = c_1 = \frac{a+b}{2}$ and $b_1 = d_1 = \sqrt{ab}$, so Borchardt's map is a generalization of the AGM map. One can prove (this is the easy part of Borchardt's work) that for any positive numbers a,b,c and d, there exists a positive number λ (dependent on a,b,c, and d) such that

$$\lim_{k\to\infty} h^k(a,b,c,d) = (\lambda,\lambda,\lambda,\lambda) \tag{1.8}$$

As a final example, generalizing the AGM map and Borchardt's map, we mention a "monster algorithm" introduced in [7]. Let G be a compact Hausdorff topological group and let μ denote Haar measure on G. Let $C(G) = X$ denote the Banach space of continuous, real-valued functions on G in the sup norm, let $L^{\frac{1}{2}}(G)$ denote the set of μ-measurable functions x such that

$$\int_G |x(s)|^{\frac{1}{2}}\, d\mu(s) < \infty$$

Let K denote the set of functions in X which are nonnegative everywhere on G and $\overset{\circ}{K}$ the functions which are everywhere strictly positive on G and define a function $F\colon K \to K$ by

$$F(x) = \sqrt{x} * \sqrt{x},$$

where $*$ denotes convolution, so

$$(\sqrt{x} * \sqrt{x})\,(t) = \int_G \sqrt{x(s)}\,\sqrt{x(ts^{-1})}\, d\mu(s).$$

If $x \in K$ it is proved in [7] that there exists a constant function $x_\infty \in K$ (x_∞ depends on x) such that

$$\lim_{k\to\infty} F^k(x) = x_\infty$$

Of course one has $x_\infty \in \overset{\circ}{K}$ if $x \in \overset{\circ}{K}$. If x is nonnegative almost everywhere, measurable and $x \in L^{\frac{1}{2}}(G)$, one can easily check that $F^2(x) \in K$, so one obtains convergence results in this case also. It is noted in [7] that the AGM map corresponds to $G = C_2$, the group with two elements; and the Borchardt map corresponds to $G = C_2 \times C_2$.

In Section 3 of [25], some general propositions are given which insure convergence of iterated means. Some of these results are summarized in [27]. In particular, Theorem 3.2 in Section 3 of [25] can be applied to all of the previously mentioned examples and yields convergence. However, because the maps we shall consider do not, in general, preserve a natural partial ordering, Theorem 3.2 from [25] will be of no help in treating extensions of the AGM to the case where the variables are linear operators.

2. GENERALIZING THE AGM FOR BOUNDED LINEAR OPERATORS

We begin by recalling some standard definitions and establishing notation. If H is a Hilbert space, $\mathscr{L}(H)$ will always denote the set of bounded linear operators from H to H. If $\langle L_k \rangle$ is a sequence of bounded linear operators in $\mathscr{L}(H)$ we shall say that L_k approaches L in the operator norm topology as $k \to \infty$ and write

$$\text{n-}\lim_{k\to\infty} L_k = L \quad \text{if and only if}$$

$$\lim_{k\to\infty} \|L_k - L\| = 0.$$

We shall say that L_k approaches L in the strong operator topology and write

$$\text{s}-\lim_{k\to\infty} L_k = L \quad \text{if and only if}$$

$$\lim_{k\to\infty} \|L_k(x) - L(x)\| = 0 \quad \text{for all } x \in H.$$

We shall say that L_k approaches L in the weak operator topology and write

$$w - \lim_{k\to\infty} L_k = L \quad \text{if and only if}$$

$$\lim_{k\to\infty} \langle L_k(x),y\rangle = \langle L(x),y\rangle \quad \text{for all } x,y \in H.$$

(Of course $\langle u,v\rangle$ denotes the inner product in H.) It is well known that $w - \lim_{k\to\infty} L_k = L$ if and only if

$$\lim_{k\to\infty} \langle L_k(x),x\rangle = \langle L(x),x\rangle \quad \text{for all } x \in H.$$

Recall, also, that if H is finite dimensional, all these notions of convergence are equivalent.

Of course an operator $A \in \mathscr{L}(H)$ is self-adjoint if $\langle Ax,y\rangle = \langle x,Ay\rangle$ for all x and $y \in H$. We shall say that an operator $B \in \mathscr{L}(H)$ is positive semi-definite if $\langle Bx,x\rangle \geq 0$ for all $x \in H$; and B is positive definite if there exists $c > 0$ such that

$$\langle Bx,x\rangle \geq c\|x\|^2 \quad \text{for all } x \in H.$$

(Some authors say "strictly positive" instead of "positive definite".) *For notational convenience we shall write that* $A \in \mathscr{L}(H)$ *is p.d. if* A *is self-adjoint and positive definite, and we shall say that* $A \in \mathscr{L}(H)$ *is p.s.d. if* A *is self-adjoint and positive semi-definite.* For the remainder of this paper, operators will almost exclusively be p.d. or p.s.d. The case of unbounded operators is also of interest (see remarks in [29]) but will not be treated here.

In generalizing the examples of Section 1 to the operator-valued case, we shall take the view that p.d. operators on a Hilbert space H are the analogues of positive reals. If A and B are p.d. and elements of $\mathscr{L}(H)$, it remains to find a reasonable analogue for expressions like $\sqrt{ab}$, where a and b are positive reals. Of course one has a functional calculus for bounded linear operators by means of Dunford's integral (see [32], p.225) and a more powerful functional calculus for self-adjoint operators (see [8,32]). One wants the analogue of $\sqrt{ab}$ to be p.d. if A and B are p.d. and to agree with $(AB)^{\frac{1}{2}}$ if A and B commute. The problem, of course, is that if A and B do not commute, AB is not self-adjoint. One might try $(\frac{1}{2}AB + \frac{1}{2}BA)^{\frac{1}{2}}$, but it is well known (see [10]) that $AB + BA$ need not be p.d. One possible solution is provided by the following simple observation, which has been noted without proof in [28] and (in less general form and also without proof) in [10].

LEMMA 2.1 Let H be Hilbert space and suppose that $A_i \in \mathscr{L}(H)$ and A_i is p.d for $1 \leq i \leq m$. Let σ_i, $1 \leq i \leq m$, be positive reals such that $\sum_{i=1}^{m} \sigma_i = 1$. For nonzero real numbers r define

$$M_{r\sigma}(A) = \left(\sum_{i=1}^{m} \sigma_i A_i^r\right)^{\frac{1}{r}}.$$

Then one has that

$$n - \lim_{r \to 0} M_{r\sigma}(A) = \exp\left(\sum_{i=1}^{m} \sigma_i \log(A_i)\right)$$

Proof. By taking logarithms it suffices to prove that

$$n - \lim_{r \to 0} r^{-1} \log\left(\sum_{i=1}^{m} \sigma_i A_i^r\right) = \sum_{i=1}^{m} \sigma_i \log(A_i). \tag{2.1}$$

We use the standard "big oh" notation, so $O(r^k)$ will denote a bounded, self-adjoint linear operator B_r such that $\|B_r\| \leq Mr^k$ for $|r|$ sufficiently small and M denotes a constant independent of r. By using the power series for the exponential we find that

$$A_i^r = \exp(r \log(A_i)) = I + r \log(A_i) + 0(r^2) \qquad (2.2)$$

It follows that

$$\sum_{i=1}^{m} \sigma_i A_i^r = I + \sum_{i=1}^{m} \sigma_i r \log(A_i) + 0(r^2). \qquad (2.3)$$

If we recall that for $X \in \mathscr{L}(H)$ with $\|X\| < 1$ we have that

$$\log(I+X) = \sum_{k=1}^{\infty} (\frac{(-1)^{k-1}}{k}) X^k,$$

we obtain from eq. (2.2) that (for $|r|$ sufficiently small)

$$\log(\sum_{i=1}^{m} \sigma_i A_i^r) = \sum_{i=1}^{m} \sigma_i r \log(A_i) + 0(r^2), \qquad (2.4)$$

and eq. (2.4) yields eq. (2.1). □

Motivated by Lemma 2.1, we now consider a generalization of eq. (1.5). For $0 < \alpha, \beta < 1$ and for p.d. operators $A, B \in \mathscr{L}(H)$, define

$$g(A,B) = (\alpha A + (1-\alpha)B, \exp(\beta \log A + (1-\beta) \log B)) \qquad (2.5)$$

We shall give a more or less self–contained proof of the following theorem, which is a very special case of results in [28].

THEOREM 2.1 (Compare [28]) Suppose that H is a Hilbert space and A and B are p.d. operators in $\mathscr{L}(H)$. For fixed numbers α,β with $0 < \alpha, \beta < 1$ define a map g by eq. (2.5). Then there exist a p.d.

operator $E \in \mathscr{L}(H)$, E dependent on A and B, such that

$$s - \lim_{k\to\infty} g^k(A,B) = (E,E) \tag{2.6}$$

Much more general theorems are proved in [28]. However, this generality also obscures some of the basic ideas. Here we shall give a proof which illustrates the basic ideas while avoiding many technical complications.

Recall that there is a partial ordering on the set A of bounded, self-adjoint linear operators in $\mathscr{L}(H)$, namely $A \leq B$ if and only if $B - A$ is p.s.d. More generally, if $(A_1, A_2,..., A_m) = A$ and $(B_1, B_2,...,B_m) = B$ are m-tuples of bounded, self-adjoint linear operators in $\mathscr{L}(H)$, we shall say that $A \leq B$ if $A_i \leq B_i$ for $1 \leq i \leq m$. If D is a convex subset of $\mathscr{L}(H)$, a map φ: $D \to A$ is called "concave" if, for all $A, B \in D$, one has

$$\varphi((1-t)A + tB) \geq (1-t)\varphi(A) + t\varphi(B) \quad \text{for} \quad 0 \leq t \leq 1.$$

The map φ is called "convex" if $-\varphi$ is concave. The following lemma is a special case of C. Loewner's beautiful classical theory: see [17] for an exposition. One can also give a relatively short, direct proof by first proving the convexity of the map $A \to A^{-1}$ for A p.d and then writing

$$\log(A) = \int_0^1 (A-I)[(1-t)I + tA]^{-1}dt.$$

LEMMA 2.2 (C.Loewner) Let H be a Hilbert space, D be the set of p.d. operators in $\mathscr{L}(H)$ and $\mathscr{A}$ be the set of self-adjoint operators in $\mathscr{L}(H)$. The map φ: $D \to \mathscr{A}$ defined by $\varphi(A) = \log(A)$ is concave.

With the aid of Lemma 2.2 we can prove a crucial lemma.

LEMMA 2.3 Let notation and hypotheses be as in Theorem 2.1 and define $(A_1,B_1) = g(A,B)$ and a matrix M by

$$M = \begin{pmatrix} \alpha & \beta \\ 1-\alpha & 1-\beta \end{pmatrix} \tag{2.7}$$

Then one has that

$$(\log A_1, \log B_1) \geq (\log A, \log B)M = (\alpha \log A + (1-\alpha)\log B, \beta \log A + (1-\beta)\log B) \tag{2.8}$$

Proof. This is immediate from Lemma 2.2. □

The matrix M is "nonnegative" in the sense that all entries of M are nonnegative, and M is "primitive," ie, there exists $p \geq 1$ such that all entries of M^p are positive (in our case, $p = 1$). Furthermore, M is "column stochastic," ie, all columns of M sum to 1. One can easily prove directly (or use the classical theory of Perron and Frobenius) that there exists a column vector $\binom{u_1}{u_2}$ with $u_1 > 0$, $u_2 > 0$ and $u_1 + u_2 = 1$, such that

$$M\binom{u_1}{u_2} = \binom{u_1}{u_2}. \tag{2.9}$$

Furthermore one has that

$$\lim_{k\to\infty} M^k = \begin{pmatrix} u_1 & u_1 \\ u_2 & u_2 \end{pmatrix}. \tag{2.10}$$

Our next two lemmas are special cases of results in Sections 1 and 2 of [28], but the arguments in our case are much simpler.

LEMMA 2.3 Let g be defined as in eq. (2.5) $(0<\alpha, \beta<1)$ and let M be as in eq. (2.7) and $u = \binom{u_1}{u_2}$ as in eq. (2.9). If A and B are p.d. operators on a Hilbert space H and

$$g^k(A,B) = (A_k, B_k),$$

there exist positive constants γ and δ such that

$$\gamma I \le A_k,\ B_k \le \delta I \quad \text{for all} \quad k \ge 1. \tag{2.11}$$

Furthermore, if $F_k = u_1 \log(A_k) + u_2 \log(B_k)$, there exists a self-adjoint bounded linear operator F such that

$$s - \lim_{k \to \infty} F_k = F \tag{2.12}$$

Proof. If E is a self-adjoint operator, it is wellknown that the spectrum of E, $\sigma(E)$, is contained in an interval $[\gamma,\delta]$ if and only if $\gamma I \le E \le \delta I$. Furthermore, the spectral mapping theorem implies that if f is a continuous, real-valued function defined on $[\gamma,\delta]$ where $\sigma(E) \subset [\gamma,\delta]$, then $\sigma(f(E)) \subset f([\gamma,\delta])$. Using these facts one sees that if γ and δ are chosen as positive numbers such that $\gamma I \le A \le \delta I$ and $\gamma I \le B \le \delta I$, then one has that $\gamma I \le A_1 \le \delta I$ and $\gamma I \le B_1 \le \delta I$. By induction one concludes that eq. (2.11) is satisfied for all $k \ge 1$. Furthermore, by using (2.11) and the same sort of argument sketched above one obtains

$$(\log\gamma)I \le F_k \le (\log\delta)I \tag{2.13}$$

Eq. (2.13) shows that F_k is bounded above, and if we can prove that the sequence F_k is monotonic increasing, the conclusion of the lemma will follow (see [8,32]). However, by the concavity of log, we obtain

$$\begin{aligned} u_1 \log(A_{k+1}) + u_2 \log(B_{k+1}) &= F_{k+1} \\ &\ge u_1(\alpha \log A_k + (1-\alpha) \log B_k) + u_2(\beta \log A_k + (1-\beta) \log B_k) \\ &= u_1 \log A_k + u_2 \log B_k = F_k, \end{aligned}$$

so the sequence F_k is monotone increasing. □

LEMMA 2.4 Let notation and assumptions be as in Lemma 2.3. Then one has

$$w - \lim_{k\to\infty} \log(A_k) = F \quad \text{and} \quad w - \lim_{k\to\infty} \log(B_k) = F \tag{2.14}$$

Proof. By using inequality (2.8) repeatedly we obtain

$$(\log A_{k+p}, \log B_{k+p}) \geq (\log A_k, \log B_k)M^p \tag{2.15}$$

For notational convenience define M_∞ by

$$M_\infty = \begin{pmatrix} u_1 & u_1 \\ u_2 & u_2 \end{pmatrix} = \lim_{p\to\infty} M^p.$$

Also, if C and D are any two bounded linear operators, define $\|(C,D)\|$ by

$$\|(C,D)\| = \sup(\|C\|, \|D\|).$$

Inequality (2.15) can be written in the form

$$\begin{aligned}(\log A_{k+p}, \log B_{k+p}) &\geq (\log A_k, \log B_k)M_\infty + (\log A_k, \log B_k)(M^p - M_\infty) \\ &= (F_k, F_k) + (\log A_k, \log B_k)(M^p - M_\infty)\end{aligned} \tag{2.16}$$

Lemma 2.3 implies that $\|\log A_k\|$ and $\|\log B_k\|$ are uniformly bounded for $k \geq 1$, so by using eq. (2.10) we see that there exists $\epsilon_p > 0$ such that for all $k \geq 1$,

$$\|(\log A_k, \log B_k)(M^p - M_\infty)\| \leq \epsilon_p, \text{ where } \lim_{p\to\infty} \epsilon_p = 0. \tag{2.17}$$

By using equations (2.12), (2.16) and (2.17), we see that for any fixed $x \in H$ and any $\epsilon > 0$ there exist N such that for all $m \geq N$ on has

$$\langle(\log A_m)x,x\rangle \geq \langle Fx,x\rangle - \epsilon \quad \text{and} \quad \langle(\log B_m)x,x\rangle \geq \langle Fx,x\rangle - \epsilon.$$

We obtain immediately from the above inequality that

$$\lim_{m\to\infty} \inf \langle (\log A_m)x,x \rangle \geq \langle Fx,x \rangle \text{ and}$$

$$\lim_{m\to\infty} \inf \langle (\log B_m)x,x \rangle \geq \langle Fx,x \rangle. \tag{2.18}$$

If, on the other hand, one has that

$$\lim_{m\to\infty} \sup \langle (\log A_m)x,x \rangle > \langle Fx,x \rangle, \tag{2.19}$$

one obtains from (2.18) and (2.19) that

$$\lim_{m\to\infty} \sup \langle (u_1 \log A_m + u_2 \log B_m)x,x \rangle > \langle Fx,x \rangle,$$

which contradicts eq. (2.12). Thus eq. (2.19) is impossible, and we must have

$$\lim_{m\to\infty} \langle (\log A_m)x,x \rangle = \langle Fx,x \rangle.$$

The same argument shows that

$$\lim_{m\to\infty} \langle (\log B_m)x,x \rangle = \langle Fx,x \rangle.$$

Since $x \in H$ was arbitrary, the lemma is proved. □

Proof of Theorem 2.1 for $\dim H < \infty$ or for $\alpha = \beta$

If H is finite dimensional, weak, strong and operator norm topologies are identical, so Lemma 2.4 implies

$$n - \lim_{k\to\infty} \log(A_k) = F \quad \text{and} \quad n - \lim_{k\to\infty} \log(B_k) = F.$$

It follows by the continuity of the exponential map in the operator norm topology that

$$n - \lim_{k\to\infty} A_k = \exp(F) = E = n - \lim_{k\to\infty} B_k.$$

The general case involves some subtleties. It is well known (see Lemma 1.2 in [28]) that if $<G_k>$ is a sequence of bounded linear operators in a Banach space and

$$s - \lim_{k\to\infty} G_k = G, \text{ then}$$

$$s - \lim_{k\to\infty} \exp(G_k) = \exp(G).$$

The analogous result for the weak operator topology is false; one can easily construct counter–examples: see Remark 1.1 in [28]. Thus to prove Theorem 2.1 it suffices to prove

$$s - \lim_{k\to\infty} \log A_k = F = s - \lim_{k\to\infty} \log B_k,$$

but one must be careful about working in the weak operator topology.

If $\alpha = \beta$, one immediately sees that $u_1 = \beta$ and $u_2 = 1-\beta$ (where u_1 and u_2 are as in eq. (2.9)). Lemma 2.3 implies that

$$s - \lim_{k\to\infty} u_1 \log A_k + u_2 \log B_k = F$$
$$= s - \lim_{k\to\infty} \beta \log A_k + (1-\beta) \log B_k = s - \lim_{k\to\infty} \log B_{k+1}.$$

It follows that

$$s - \lim_{k\to\infty} B_k = \exp(F) = E$$

If δ is as in eq. (2.11) we know that

$$\|A_k\| \le \delta, \ \|B_k\| \le \delta \ \text{ and } \ \|E\| \le \delta \ \text{ for } \ k \ge 1.$$

By repeated application of the defining equation

$$A_{k+1} = \alpha A_k + (1-\alpha)B_k,$$

we obtain (for $p \ge 1$)

$$A_{k+p} = \alpha^p A_k + (1-\alpha)\sum_{j=1}^{p} \alpha^{j-1} B_{k+p-j}. \tag{2.20}$$

By using eq. (2.20) we see that

$$A_{k+p} - E = \alpha^p(A_k - E) + (1-\alpha)\sum_{j=1}^{p} \alpha^{j-1}(B_{k+p-j} - E). \tag{2.21}$$

For a fixed $x \in H$ and $\epsilon > o$, select N_1 so large that

$$\|B_i x - Ex\| < \frac{\epsilon}{2} \ \text{ for } \ i \ge N_1.$$

If $k \ge N_1$, eq. (2.21) then yields

$$\|A_{k+p}x - Ex\| \le \alpha^p(2\gamma)\|x\| + (1-\alpha)\sum_{j=1}^{p} \alpha^{j-1}\left(\frac{\epsilon}{2}\right)$$

$$\le 2\alpha^p\gamma\|x\| + \left(\frac{\epsilon}{2}\right)(1-\alpha^p) \tag{2.22}$$

It follows from (2.22) that there exists N_2 such that for $m \ge N_2$,

$$\|A_m x - Ex\| < \epsilon \qquad \square$$

If one knows that $u_1\log(A_k) + u_2\log(B_k)$ converges to F in operator norm and if $\alpha = \beta$, then a slight modification of the above argument shows that

$$\lim_{k\to\infty} \|A_k - E\| = \lim_{k\to\infty} \|B_k - E\| = 0. \tag{2.23}$$

In general, one can ask whether eq. (2.23) is satisfied whenever $\alpha = \beta$ and A and B are p.d. operators on an infinite dimensional Hilbert space. The answer is not known.

It remains to consider Theorem 2.1 in the case that H is infinite dimensional and $\alpha \neq \beta$. To handle this case we seem to need more than just concavity of the map $A \to \log(A)$. Rather it seems necessary to exploit what could be called "uniform concavity" of the map $A \to \log(A)$. To be precise we need the following theorem from [28].

THEOREM 2.2 (See Theorem 1.2 in [28]) Let H be a Hilbert space and suppose that $\langle A^{(k)} \rangle$, $k \geq 1$, is a sequence of m–tuples of p.d. operators on H, so

$$A^{(k)} = (A_1^{(k)}, A_2^{(k)}, \ldots, A_m^{(k)})$$

and $A_j^{(k)}$ is a p.d. operator on H. Assume that θ: $(0,\infty) \to \mathbb{R}$ is a C^1, real–valued function such that $\lim_{x\to\infty} \theta'(x) = 0$ and such that θ has an analytic extension to $U = \{z \in \mathbb{C}\colon \operatorname{Im}(z) \neq 0$ or $\operatorname{Im}(z) = 0$ and $\operatorname{Re}(z) > 0\}$ for which $\operatorname{Im}(\theta(z)) > 0$ for all z such that $\operatorname{Im}(z) > 0$. Assume that there exist fixed positive numbers λ_p, $1 \leq p \leq m$, such that $\sum_{p=1}^{m} \lambda_p = 1$ and

$$w - \lim_{k\to\infty} [\theta(\sum_{p=1}^{m} \lambda_p A_p^{(k)}) - \sum_{p=1}^{m} \lambda_p \theta(A_p^{(k)})] = 0.$$

Assume also that there exist fixed positive numbers α and β such that

$$\alpha I \leq A_p^{(k)} \leq \beta I \quad \text{for } k \geq 1 \text{ and } 1 \leq p \leq m.$$

Then for any i and j with $1 \leq i, j \leq m$ one has

$$s - \lim_{k\to\infty} [\theta(A_i^{(k)}) - \theta(A_j^{(k)})] = 0.$$

Notice that $\theta(x) = \log(x)$, $\theta(x) = -x^{-1}$ and $\theta(x) = x^{\gamma}$, $0 < \gamma < 1$, all satisfy the hypotheses of Theorem 2.2.

Proof of Theorem 2.1 (assuming Theorem 2.2)

The idea is to apply Theorem 2.2 to the sequence of 2–tuples (A_k,B_k), $k \geq 1$ and with $\theta = \log$. Lemma 2.4 implies that

$$w - \lim_{k\to\infty} \theta(A_{k+1}) = w - \lim_{k\to\infty} \theta(A_k) = w - \lim_{k\to\infty} \theta(B_k) = F, \text{ so}$$

$$w - \lim_{k\to\infty} [\log(\alpha A_k + (1-\alpha)B_k) - \alpha \log A_k - (1-\alpha)\log B_k] = 0.$$

Lemma 2.3 implies that there exist $\gamma, \delta > 0$ such that

$$\gamma I \leq A_k, B_k \leq \delta I \quad \text{for} \quad k \geq 1.$$

Theorem 2.2 thus implies (take $\lambda_1 = \alpha$ and $\lambda_2 = 1 - \alpha$) that

$$s - \lim_{k\to\infty}(\log(A_k) - \log(B_k)) = 0. \tag{2.24}$$

Because we know that

$$s - \lim_{k\to\infty} u_1 \log A_k + u_2 \log B_k = F, \tag{2.25}$$

we conclude from (2.24) and (2.25) that

$$s - \lim_{k\to\infty} \log A_k = s - \lim_{k\to\infty} \log B_k = F$$

and the theorem is proved. □

3. ALTERNATE DEFINITIONS OF THE AGM FOR OPERATORS

If A and B are p.s.d. operators on a Hilbert space H, Pusz and Woronowicz [29] have defined a p.s.d. operator, which (following Ando [4,5] and Trapp [30]) we shall denote $A \# B$, and which can be considered an analogue of the geometric mean of two positive reals. The basic properties of $A \# B$ have been developed by Pusz–Woronowicz [29], but Ando [4,5] has given a development which is somewhat more suitable for our needs. We list below some useful properties of this operation:

(a) If A and B are p.d. operators on H, then one has

$$A \# B = A^{\frac{1}{2}}(A^{-\frac{1}{2}}BA^{-\frac{1}{2}})^{\frac{1}{2}} A^{\frac{1}{2}}. \tag{3.1}$$

(b) If A, B and C are p.s.d. operators on H, then $L = \begin{bmatrix} A & C \\ C & B \end{bmatrix}$ defines a bounded, self–adjoint linear operator on $H \oplus H$ by

$$L\begin{pmatrix} x_1 \\ x_2 \end{pmatrix} = \begin{pmatrix} Ax_1 + Cx_2 \\ Cx_1 + Bx_2 \end{pmatrix}.$$

If $C = A \# B$, then L is positive semi–definite, and if D is any p.s.d. operator on H such that $\begin{bmatrix} A & D \\ D & B \end{bmatrix}$ is p.s.d., then $D \leq A \# B$, so $A \# B$ is the maximal p.s.d. operator C such that $\begin{bmatrix} A & C \\ C & B \end{bmatrix}$ is p.s.d.

As an immediate consequence of property (b) one obtains the following property:

(c) If $0 \leq A_1 \leq A_2$ and $0 \leq B_1 \leq B_2$ (where the A_j and B_j are p.s.d. operators on H), then $A_1 \# B_1 \leq A_2 \# B_2$.

(d) Suppose that A_k is a monotonic decreasing sequence of p.s.d. operators such that $\text{s-}\lim_{k \to \infty} A_k = A$ and B_k is a monotonic decreasing sequence of p.s.d. operators such that

$$s - \lim_{k\to\infty} B_k = B. \quad \text{If} \quad C_k = A_k \# B_k,$$
$$s - \lim_{k\to\infty} C_k = A \# B.$$

Proof. By using property (c) we see that C_k is a monotonic decreasing sequence of p.s.d. operators, so

$$s - \lim_{k\to\infty} C_k = C \quad \text{exists.}$$

Since $\begin{bmatrix} A_k & C_k \\ C_k & B_k \end{bmatrix}$ is p.s.d., it follows by definition of the inner product on $H \oplus H$ that for any $u, v \in H$ we have

$$\langle A_k u,u\rangle + \langle C_k v,u\rangle + \langle C_k u,v\rangle + \langle B_k v,v\rangle \geq 0.$$

Taking limits gives

$$\langle Au,u\rangle + \langle Cv,u\rangle + \langle Cu,v\rangle + \langle Bv,v\rangle \geq 0,$$

so that $\begin{bmatrix} A & C \\ C & B \end{bmatrix}$ is p.s.d.

If D is p.s.d. and $\begin{bmatrix} A & D \\ D & B \end{bmatrix}$ is p.s.d., it follows (because $A \leq A_k$ and $B \leq B_k$) that $\begin{bmatrix} A_k & D \\ D & B_k \end{bmatrix}$ is p.s.d. and that (because of property (b)) $D \leq C_k$. Taking limits gives $D \leq C$, so C is the maximal p.s.d. operator for which $\begin{bmatrix} A & C \\ C & B \end{bmatrix}$ is p.s.d. and therefore (property (b)) $C = A \# B$.

If one assumes that A and B are p.d. and $A \# B$ is given by the formula in (a), then one can prove (see [5]) that $A \# B$ satisfies property (b) and therefore property (c). If A and B are just p.s.d. one can find monotonic decreasing sequences of p.d. operators A_k and B_k such that

$$s - \lim_{k\to\infty} A_k = A \quad \text{and} \quad s - \lim_{k\to\infty} B_k = B$$

and one can *define* A # B by

$$A \# B = s - \lim_{k\to\infty} A_k \# B_k.$$

It is not hard to prove that this definition is independent of the particular sequences and that property (b) (and hence also properties (c) and (d)) are still satisfied with this extended definition of A # B.

Anderson and Trapp [3] have observed that

$$A \# B = A(A^{-1}B)^{\frac{1}{2}}. \tag{3.2}$$

More generally, it is proved in Section 4 of [28] that if A and B are p.d. operators on a Hilbert space H and λ is any real number then

$$B \# A = B^{\lambda}(B^{-\lambda}AB^{\lambda-1})^{\frac{1}{2}}\, B^{1-\lambda} = B(B^{-1}A)^{\frac{1}{2}} \text{ and} \tag{3.3}$$

$$A \# B = A^{\lambda}(A^{-\lambda}BA^{\lambda-1})^{\frac{1}{2}}\, A^{1-\lambda} = A(A^{-1}B)^{\frac{1}{2}}. \tag{3.4}$$

Of course it is also well known that

$$A \# B = B \# A.$$

Some explanation of equations (3.3) and (3.4) is needed. For λ a real number and $z \neq 0$ a complex number which is not a negative real, we shall always use the standard single–valued branch of z^{λ} and log(z). Equations (3.3) and (3.4) are to be interpreted using Dunford's functional calculus [32]. We know $B^{-\frac{1}{2}}AB^{-\frac{1}{2}}$ is p.d. and hence has spectrum contained in the positive reals. However, we can write (using σ to denote the spectrum of an operator)

$$B^{-\lambda}AB^{\lambda-1} = S(B^{-\frac{1}{2}}AB^{-\frac{1}{2}})S^{-1},\ S = B^{-\lambda+\frac{1}{2}}, \text{ so}$$

$$\sigma(B^{-\lambda}AB^{\lambda-1}) = \sigma(B^{-\frac{1}{2}}AB^{-\frac{1}{2}}) \subset (0,\infty),$$

and eq. (3.3) makes sense via Dunford's functional calculus.

If A, B and C are elements of $\mathscr{L}(H)$ and are p.s.d., one can easily prove that $\begin{bmatrix} A & C \\ C & B \end{bmatrix}$ is p.s.d. if and only if

$$\langle Ax,x\rangle\langle By,y\rangle \geq |\langle Cx,y\rangle|^2 \text{ for all } x,y \in H. \tag{3.6}$$

If N(D) denotes the null space of $D \in \mathscr{L}(H)$ and range (D) denotes the range, one derives from eq. (3.6) that if A, B and C are p.s.d. and $\begin{bmatrix} A & C \\ C & B \end{bmatrix}$ is p.s.d., than $N(A) \subset N(C)$ and $N(B) \subset N(C)$. Furthermore, one can derive from eq. (3.6) by taking $y = C(x)$ that

$$\|B\| \langle (A^{\frac{1}{2}})^2 x,x\rangle \geq \langle C^2 x,x\rangle. \tag{3.7}$$

Equation (3.7) and Theorem 1 in [18] then imply that

$$\text{range}(C) \subset \text{range}(A^{\frac{1}{2}}),$$

and a similar argument shows that

$$\text{range}(C) \subset \text{range}(B^{\frac{1}{2}}).$$

Thus we have proved that

(e) If A and B are p.s.d. operators in $\mathscr{L}(H)$, one has

$$N(A) \subset N(A\#B) \text{ and } N(B) \subset N(A\#B) \text{ and} \tag{3.8}$$

$$\text{range}(A\#B) \subset \text{range}(A^{\frac{1}{2}}) \cap \text{range}(B^{\frac{1}{2}}). \tag{3.9}$$

Property (e) also follows directly from Theorem I.1 in [4].

On the basis of property (d) one might expect that the map

$(A,B) \to A \# B$ is continuous in the strong operator topology on $K \times K$, where K is the set of p.s.d. operators in $\mathscr{L}(H)$. Indeed, it is easy to see that if $A_k, B_k \in K$ and

$$w - \lim_{k\to\infty} A_k = A, \; w - \lim_{k\to\infty} B_k = B \text{ and } w - \lim_{k\to\infty} A_k \# B_k = C,$$

then $C \leq A \# B$

However, by using property (e) one can easily construct examples which show that $(A,B) \to A \# B$ is *not* continuous on $K \times K$ even when H is two dimensional.

With these preliminaries we can return to our discussion of the AGM. If $C = K \times K$ (K as above), J. Fujii [19] has defined a map $f\colon C \to C$ by

$$f(A,B) = (\tfrac{A+B}{2}, A \# B) \tag{3.10}$$

For any $(A,B) \in C$, Fujii has proved that

$$s - \lim_{k\to\infty} f^k(A,B) = (E,E), \; E \in K. \tag{3.11}$$

If A and B are positive definite, it is proved in Section 4 of [28] that E is p.d. and that convergence in (3.11) is in the operator norm.

For $\lambda \geq 1$ one can define a map $f_\lambda\colon C \to C$ by

$$f_\lambda(A,B) = (\tfrac{A+B}{2}, (A^{\frac{1}{\lambda}} \# B^{\frac{1}{\lambda}})^\lambda). \tag{3.12}$$

For each $\lambda \geq 1$ the map f_λ can be considered an analogue of AGM map for positive reals. It is proved in Theorem 4.3 of [28] that if A and B are p.d., then

$$n-\lim_{\lambda\to\infty}\left(A^{\frac{1}{\lambda}}\ \#\ B^{\frac{1}{\lambda}}\right)^{\lambda} = \exp\left(\tfrac{1}{2}\log A + \tfrac{1}{2}\log B\right) \tag{3.13}$$

Thus, for $\lambda = 1$ one obtains Fujii's map and for $\lambda = \infty$ one obtains the map in Section 2. It is also proved in Theorem 4.2 of [28] that if A and B are p.d. and $\lambda \geq 1$, then

$$s-\lim_{k\to\infty} f_{\lambda}^{k}(A,B) = (E_{\lambda},E_{\lambda}) \tag{3.14}$$

where E_{λ} is p.d. However, the argument in Theorem 4.2 of [28] does not directly extend to the case that A and B are p.s.d. instead of p.d, and we wish to handle the positive semi-definite case. We begin with a lemma.

LEMMA 3.1 Let H be a Hilbert space and $A_k \in \mathscr{L}(H)$, $k \geq 1$, a sequence of bounded, self-adjoint linear operators such that

$$s-\lim_{k\to\infty} A_k = A \quad \text{and}$$

$$\sigma(A_k) \subset [\alpha,\beta] \quad \text{for all} \quad k \geq 1.$$

If h is a continuous real-valued function defined on $[\alpha,\beta]$, then

$$s-\lim_{k\to\infty} h(A_k) = h(A).$$

Proof. By the uniform boundedness principle we know that $\|A_k\|$ is uniformly bounded, so we can assume α and β finite. Standard theory of self-adjoint operators implies that $\sigma(A) \subset [\alpha,\beta]$: see [8,32]. Lemma 3.1 is well known if h is a polynomial: see, for example, Lemma 1.2 in [28]. In general, given $\epsilon > 0$, find a real-valued polynomial p such that

$$\sup_{x\in[\alpha,\beta]} |h(x) - p(x)| < \epsilon$$

and use the fact that for any bounded, self-adjoint linear operator B with $\sigma(B) \subset [\alpha,\beta]$ one has

$$\|h(B) - p(B)\| \leq \sup\{|h(x) - p(x)| : x \in [\alpha,\beta]\}. \qquad \square$$

THEOREM 3.1 (Compare Theorem 4.2 in [28]) For $\lambda \geq 1$ let f_λ be defined by eq. (3.12) and let f_λ^k denote the kth iterate of f_λ. If H is a Hilbert space and A and B are elements of $\mathscr{L}(H)$ and are p.s.d., then there exists $E_\lambda \in \mathscr{L}(H)$, E_λ p.s.d., such that

$$s - \lim_{k\to\infty} f_\lambda^k(A,B) = (E_\lambda, E_\lambda).$$

Proof. Define maps g_λ: $C \to C$, ψ_λ: $C \to C$ and ψ_λ^{-1}: $C \to C$ by

$$g_\lambda(A,B) = \left(\left(\frac{A^\lambda + B^\lambda}{2}\right)^{\frac{1}{\lambda}}, A \# B\right)$$

$$\psi_\lambda(A,B) = \left(A^{\frac{1}{\lambda}}, B^{\frac{1}{\lambda}}\right), \ \psi_\lambda^{-1}(A,B) = (A^\lambda, B^\lambda).$$

One can easily check that

$$f_\lambda = \psi_\lambda^{-1}\, g_\lambda\, \psi_\lambda \quad \text{and} \quad f_\lambda^k = \psi_\lambda^{-1}\, g_\lambda^k\, \psi_\lambda \tag{3.15}$$

By using eq. (3.15) and Lemma 3.1 one can see that it suffices to prove that for any p.s.d. operators A and B there exists an operator G_λ such that

$$s - \lim_{k\to\infty} g_\lambda^k(A,B) \equiv s - \lim_{k\to\infty} (A_k,B_k) = (G_\lambda,G_\lambda)$$

It is also not hard to see (use property (c) and induction) that if $A \leq \beta I$

and $B \leq \beta I$, then

$$A_k \leq \beta I \text{ and } B_k \leq \beta I \text{ for all } k \geq 1. \tag{3.16}$$

Recall (see [30]) that for $\lambda \geq 1$ and A and B p.s.d. operators on H one has

$$\left(\frac{A^\lambda + B^\lambda}{2}\right)^{\frac{1}{\lambda}} \geq \left(\frac{A+B}{2}\right) \geq A \# B. \tag{3.17}$$

Using (3.17) we see that if $(A_k,B_k) = g_\lambda^k\,(A,B)$, then

$$A_k \geq B_k \text{ for } k \geq 1, \tag{3.18}$$

and (3.18) and property (c) imply that for $k \geq 1$,

$$B_{k+1} = A_k \# B_k \geq B_k \# B_k = B_k \tag{3.19}$$

Using (3.16) and (3.19) we conclude that there exists a p.s.d. operator $G_\lambda \leq \beta I$ such that

$$s - \lim_{k\to\infty} B_k = G_\lambda \equiv G$$

Lemma 3.1 implies that

$$s - \lim_{k\to\infty} B_k^\lambda = G^\lambda \tag{3.20}$$

The defining equation for A_k gives

$$A_{k+1}^\lambda = \left(\tfrac{1}{2}\right) A_k^\lambda + \left(\tfrac{1}{2}\right) B_k^\lambda \tag{3.21}$$

Using eq. (3.21) repeatedly we obtain for $m \geq 1$

$$A^{\lambda}_{k+m} = 2^{-m} A^{\lambda}_{k} + \sum_{j=1}^{m} 2^{-j} B^{\lambda}_{k+m-j} \tag{3.22}$$

Given $\epsilon > 0$ and $x \in H$, eq. (3.20) implies that there exists n such that

$$\|B^{\lambda}_{i} x - G^{\lambda} x\| < (\tfrac{\epsilon}{2}) \quad \text{for} \quad i \geq n \tag{3.23}$$

Also, because $\|A^{\lambda}_{k}\| \leq \beta^{\lambda}$, we can select n so large that

$$\|2^{-m} A^{\lambda}_{k} x\| \leq 2^{-m} \beta^{\lambda} \|x\| < \tfrac{\epsilon}{4} \quad \text{for} \quad k \geq n \quad \text{and} \quad m \geq n. \tag{3.24}$$

If $i \geq 2n$ and we write $i = k + m$ with $k \geq n$ and $m \geq n$, we obtain from (3.22)–(3.24) that

$$\|A^{\lambda}_{i} x - G^{\lambda} x\| \leq \|2^{-m}(A^{\lambda}_{k} x - G^{\lambda} x)\| + \sum_{j=1}^{m} 2^{-j} \|B^{\lambda}_{k+m-j} x - G^{\lambda} x\|$$

$$\leq \tfrac{\epsilon}{4} + \tfrac{\epsilon}{4} + \sum_{j=1}^{m} 2^{-j} (\tfrac{\epsilon}{2}) = \epsilon \tag{3.25}$$

Equation (3.25) and Lemma 3.1 imply that

$$s - \lim_{i \to \infty} A^{\lambda}_{i} = G^{\lambda} \quad \text{and} \quad s - \lim_{i \to \infty} A_{i} = G. \qquad \square$$

Theorem 3.1 is of little interest unless one can prove that in general, for $1 \leq \lambda < \mu$, one has

$$(A^{\frac{1}{\lambda}} \# B^{\frac{1}{\lambda}})^{\lambda} \neq (A^{\frac{1}{\mu}} \# B^{\frac{1}{\mu}})^{\mu} \quad \text{and} \tag{3.26}$$

$$E_{\lambda} \neq E_{\mu} \tag{3.27}$$

Of course, if $AB = BA$ we know that equality holds in equations (3.26) and (3.27). It is claimed in Section 4 of [28] (see equations (4.49–(4.51)) that, in general, inequality does hold in (3.26) and (3.27), but no proof is given. We shall prove this result here. The idea of the proof will be to note that $(A,B) \to f_\lambda(A,B)$ can be defined and is an analytic function for arbitrary $A = I + C$ and $B = I + D$ in $\mathscr{L}(H)$ such that $\|C\|$ and $\|D\|$ are small. Furthermore, for such A,B we shall show that

$$n - \lim_{k\to\infty} f_\lambda^k (A,B) = E_\lambda$$

exists and that E_λ is an analytic function of (A,B). We shall compute the first three terms of the Taylor series for E_λ in powers of C and D and prove that the cubic terms are, in general, different for different λ.

We shall need the definition and basic properties of an analytic map f defined on an open subset G of a complex Banach space Y and having range in a complex Banach space Z. Recall that f is called complex analytic on G if f is continuous and for every $y_0 \in G$, every $y \in Y$ and every complex linear functional $\psi \in Z^*$, the map

$$z \to \psi(f(y_0 + zy))$$

is complex analytic on

$$G_{y_0,y} = \{z \in \mathbb{C}: \ y_0 + zy \in G\}.$$

The basic properties of analytic maps and differential calculus in Banach spaces can be found in [16].

It will be convenient also to recall a special case of a lemma from [28]. If X is a complex Banach space, define $Y = X \times X$ and define

$$\|y\| = \|(x_1,x_2)\| = \max(\|x_1\|, \|x_2\|).$$

For $y = (x_1,x_2) \in Y$ define $p(y)$ by

$$p(y) = \|x_1-x_2\|. \tag{3.28}$$

and define $S \subset Y$ by

$$S = \{(x,x):\ x \in X\} = \{y \in Y:\ p(y) = 0\} \tag{3.29}$$

For $r > 0$ and $y \in Y$ we shall always write

$$B_r(y) = \{w \in Y:\ \|w-y\| < r\}$$

LEMMA 3.2 (See Lemma 3.1 in [28]) Let X,Y, p and S be as above. Suppose that $y^0 \in S$, that $\delta > 0$, and that $f\colon U = B_\delta(y^0) \to Y$ is a complex analytic map such that $f(y) = y$ for all $y \in S \cap U$. Assume that there exist a constant $c < 1$ and a constant k such that

$$p(f(y)) \le cp(y) \quad \text{and} \tag{3.30}$$

$$\|f(y) - y\| \le kp(y) \tag{3.31}$$

for all $y \in U$. Then there exists $r > 0$ (dependent only on c, k and δ) such that for all $y \in \overline{B_r(y^0)} = \overline{V}$, $f^m(y) \in U$ for all $m \ge 1$, $f^m(y)$ converges uniformly to $g(y)$ for $y \in \overline{V}$ as $m \to \infty$ and $f(g(y)) = g(y)$. The map $y \to g(y)$ is complex analytic on $V = B_r(y^0)$.

It will be convenient to give a slightly different form of Lemma 3.2.

LEMMA 3.3 Let notation and assumptions be as in Lemma 3.2. Assume that $R\colon U \to Y$ is a complex analytic map such that $R(y) = 0$ for all $y \in S \cap U$, $R'(y^0) = 0$ and $\|R(y)\| \le M$ for all $y \in U$. Define $\varphi(y) = f(y) + R(y)$. Then there exists $r > 0$ (dependent only on c, k, δ

and M) such that for all $y \in \overline{B_r(y^o)} = \overline{V}$, $\varphi^m(y) \in U$ for all $m \geq 1$, $\varphi^m(y)$ converges uniformly to $\gamma(y)$ for $y \in \overline{V}$ as $m \to \infty$ and $\varphi(\gamma(y)) = \gamma(y)$. The map $y \to \gamma(y)$ is analytic on V.

Proof. By Lemma 3.2, it suffices to find positive numbers δ_1, $c_1 < 1$, and k_1, all dependent only on k, c, δ and M, such that

$$p(\varphi(y)) \leq c_1 p(y) \quad \text{and} \tag{3.32}$$

$$\|\varphi(y) - y\| \leq k_1 p(y) \tag{3.33}$$

for $y \in U_1 = \overline{B_{\delta_1}(y^o)}$. If we define $c_1 = \frac{1+c}{2}$ and $\eta = \frac{1-c}{2}$ and observe that

$$p(\varphi(y)) \leq p(f(y)) + p(R(y)) \leq cp(y) + p(R(y)),$$

we see that to verify (3.32) it suffices to prove that there exists $\delta_1 > 0$ (δ_1 dependent only on c, δ, k and M) such that

$$p(R(y)) \leq \eta p(y) \quad \text{for} \quad y \in U_1.$$

If we write $R(y) = (R_1(y), R_2(y))$ and, for $y = (u,v) \in U$, define y^* by $y^* = (\frac{u+v}{2}, \frac{u+v}{2}) \in S \cap U$, we see that

$$p(R(y)) = \|R_1(y) - R_1(y^*) - R_2(y) + R_2(y^*)\|$$

$$\leq \|R_1(y) - R_1(y^*)\| + \|R_2(y) - R_2(y^*)\|.$$

Thus, to verify (3.32), it suffices to prove that, for $j = 1$ or $j = 2$,

$$\|R_j(y) - R_j(y^*)\| \le (\tfrac{\eta}{2})\ p(y) \quad \text{for all} \quad y \in U_1 \tag{3.34}$$

Of course, in deriving (3.34), we have used the fact that $R_j(y^*) = 0$.

If inequality (3.34) holds we also obtain that

$$\|\varphi(y) - y\| \le \|f(y) - y\| + \|R(y)\| \le kp(y) + (\tfrac{\eta}{2})\ p(y)$$

for all $y \in U_1$. Thus it suffices to verify (3.34).

Recall that all higher derivatives $R^{(m)}(y)$ of the analytic function $R(y)$ can be estimated on U_1 in terms of M, δ_1 and δ. Specifically, we have

$$\|R^{(m)}(y)\| \le \frac{m!M}{(\delta-\delta_1)^j} \quad \text{for all} \quad y \in U_1.$$

It follows that for $y \in U_1$ we have (writing $R' = R^{(1)}$ and $y_t = y^* + t(y-y^*)$)

$$R_j(y) - R_j(y^*) = \int_0^1 R'(y_t)(y-y^*)dt \quad \text{and}$$

$$\|R_j(y) - R_j(y^*)\| \le \|y-y^*\| \int_0^1 \|R_j'(y_t)\|dt = p(y) \int_0^1 \|R_j'(y_t)\|dt. \tag{3.35}$$

However, for $u \in U_1$, we have (writing $u_s = y^0 + s(u-y^0)$)

$$R_j'(u) - R_j'(y^0) = R_j'(u) = \int_0^1 R_j^{(2)}(u_s)(u-y^0)ds, \text{ so}$$

$$\|R_j'(u)\| \le \frac{2M}{(\delta-\delta_1)^2}\|u-y_0\| \le \frac{2M}{(\delta-\delta_1)^2}\ \delta_1 \tag{3.36}$$

Combining (3.35) and (3.36) we obtain that

$$\|R_j(y) - R_j(y^*)\| \leq \frac{2M\delta_1}{(\delta-\delta_1)^2}\, p(y) \quad \text{for all} \quad y \in U_1 \tag{3.37}$$

We see from (3.36) that by selecting δ_1 sufficiently small (say $\delta_1 \leq \frac{\delta}{2}$ and $\delta_1 \leq \frac{\eta\delta^2}{16M}$) inequality (3.34) will be satisfied. □

Our idea is to apply Lemmas 3.2 and 3.3 to the map f_λ in eq. (3.12). First we need another lemma.

LEMMA 3.4 Let H be a Hilbert space and suppose that $A = I + E \in \mathscr{L}(H)$ and $B = I + F \in \mathscr{L}(H)$. There exists a positive number δ ($\delta = \frac{1}{4}$ will work) such that if $\|E\| < \delta$ and $\|F\| < \delta$, then the map

$$(E,F) \to B(B^{-1}A)^{\frac{1}{2}}$$

is a well-defined analytic function. (Here the usual single-valued branch of $\sqrt{z}$ is taken). Furthermore, one has

$$B(B^{-1}A)^{\frac{1}{2}} = I + \tfrac{1}{2}(E+F) - (\tfrac{1}{8})(E-F)^2 + \tfrac{1}{16}(E-F)(E+F)(E-F) + R(E,F). \tag{3.38}$$

The map $(E,F) \to R(E,F)$ is an analytic function and there exists a constant M (dependent only on δ) such that

$$\|R(E,F)\| \leq M(\|E\|^2 + \|F\|^2)^2 \tag{3.39}$$

Proof. For notational convenience we shall write

$$\|(E,F)\| = \max(\|E\|, \|F\|) < \delta,$$

and we shall determine how small δ must be. Assuming that $\delta < 1$ we have

$$(I+F)^{-1}(I+E) = I + (I+F)^{-1}(E-F) = I + W.$$

We know that

$$\|(I+F)^{-1}\| = \|\sum_{k=0}^{\infty} (-1)^k F^k\| \leq (1-\|F\|)^{-1} \leq (1-\delta)^{-1}, \text{ so} \tag{3.40}$$

$$\|W\| \leq (\|F\| + \|E\|)(1-\delta)^{-1}. \tag{3.41}$$

It follows that if $\delta < \frac{1}{3}$, $\|W\| < 1$ and the binomial theorem yields

$$((I+F)^{-1}(I+E))^{\frac{1}{2}} = I + \sum_{k=1}^{\infty} \binom{\frac{1}{2}}{k} [(I+F)^{-1}(E-F)]^k \tag{3.42}$$

We obtain from (3.42) that

$$\begin{aligned}(I+F)((I+F)^{-1}(I+E))^{\frac{1}{2}} &= I + \frac{E+F}{2} - \frac{1}{8}(E-F)(I+F)^{-1}(E-F) \\ &+ \frac{1}{16}(E-F)(I+F)^{-1}(E-F)(I+F)^{-1}(E-F) \\ &+ (I+F)\sum_{k=4}^{\infty} \binom{\frac{1}{2}}{k} [(I+F)^{-1}(E-F)]^k.\end{aligned} \tag{3.43}$$

It is easy to show that

$$\left|\binom{\frac{1}{2}}{k}\right| \leq \frac{1}{4k} \text{ for } k \geq 2,$$

so, if we assume henceforth that $\delta = \frac{1}{4}$, we obtain

$$\|(I+F)\sum_{k=4}^{\infty}\binom{\frac{1}{2}}{k}[(I+F)^{-1}(E-F)]^k\| \leq$$

$$(1+\delta)(\tfrac{1}{16})\sum_{k=4}^{\infty}(\|E\| + \|F\|)^k(1-\delta)^{-k} = \tag{3.44}$$

$$(\tfrac{1+\delta}{16})(\|E\| + \|F\|)^4[1 - (\|E\| + \|F\|)(1-\delta)^{-1}]^{-1} \leq \tfrac{1}{4}(\|E\| + \|F\|)^4.$$

Using the power series for $(I+F)^{-1}$ we obtain

$$-\tfrac{1}{8}(E-F)(I+F)^{-1}(E-F) = -\tfrac{1}{8}(E-F)^2 + \tfrac{1}{8}(E-F)F(E-F)$$

$$-\tfrac{1}{8}(E-F)(\sum_{k=2}^{\infty}(-1)^kF^k)(E-F), \text{ and} \tag{3.45}$$

$$\|-\tfrac{1}{8}(E-F)(\sum_{k=2}^{\infty}(-1)^kF^k)(E-F)\| \leq \tfrac{1}{8}(\|E\|+\|F\|)^2(\tfrac{\|F\|^2}{1-\delta}) \tag{3.46}$$

Similarly, we find that

$$\tfrac{1}{16}(E-F)(I+F)^{-1}(E-F)(I+F)^{-1}(E-F) = \tfrac{1}{16}(E-F)^3 + R_3(E,F), \text{ where} \tag{3.47}$$

$$\|R_3(E,F)\| \leq (\tfrac{1}{8})(1-\delta)^{-2}\|F\|(\|F\|+\|E\|)^3 \leq \tfrac{2}{9}\|F\|(\|F\|+\|E\|)^3.$$

Combining equations (3.43)–(3.47) we obtain eq. (3.38) and we find that for $\delta = \frac{1}{4}$ and $\|(E,F)\| < \delta$

$$\|R(E,F)\| \leq (\tfrac{1}{4})(\|E\|+\|F\|)^4 + (\tfrac{1}{6})(\|E\|+\|F\|)^2\|F\|^2 + (\tfrac{2}{9})\|F\|(\|F\|+\|E\|)^3.$$

□

Notice that by using Lemma 3.4 and decreasing δ we can assume that $B(B^{-1}A)^{\frac{1}{2}} = I+W$ where $\|W\| < 1$.

In stating our next theorem we use the standard Lie bracket notation:

$$[A,B] = AB-BA \quad \text{for} \quad A, B \in \mathscr{L}(H).$$

THEOREM 3.2 Let H be a complex Hilbert space, $\lambda \geq 1$ a real number and $\mu = \lambda^{-1}$. There exists $\epsilon > 0$, ϵ independent of $\lambda \geq 1$, such that if

$$U = \{(A,B): \ A, B \in \mathscr{L}(H), \ A = I+C, \ B = I+D \ \text{ and } \ \|(C,D)\| < \epsilon\},$$

the map $(A,B) \to f_\lambda(A,B)$ given by

$$f_\lambda(A,B) = (\tfrac{A+B}{2}, \ [B^\mu(B^{-\mu}A^\mu)^{\frac{1}{2}}]^\lambda)$$

is analytic. If $E_1 \in \mathscr{L}(H)$ is defined by

$$E_1 = (\tfrac{1}{24})\{[C,D]C + C[D,C] + D[C,D] + [D,C]D\},$$

one has the following equation for $(A,B) \in U$:

$$[B^\mu(B^{-\mu}A^\mu)^{\frac{1}{2}}]^\lambda = I + (\tfrac{1}{2})(C+D) - (\tfrac{1}{8})(C-D)^2 + (\tfrac{1}{16})(C-D)(C+D)(C-D)$$

$$+ (\mu-1)E_1 + (\tfrac{1}{2})(\mu-1)^2E_1 + R_4(\mu,C,D). \tag{3.48}$$

There exists a constant M, independent of $\lambda \geq 1$, such that for $\|(C,D)\| < \epsilon$ one has

$$\|R_4(\mu,C,D)\| \leq M \ (\|C\|^2 + \|D\|^2)^2 \tag{3.49}$$

Proof. For $\lambda \geq 1$ and $\|(C,D)\| < \epsilon < 1$, the binomial theorem gives

$$(I+C)^{\mu} = I+E = I+\sum_{k=1}^{\infty} \binom{\mu}{k} C^k \quad \text{and} \tag{3.50}$$

$$(I+D)^{\mu} = I+F = I + \sum_{k=1}^{\infty} \binom{\mu}{k} D^k \tag{3.51}$$

Because $0 < \mu \leq 1$, one easily sees that

$$\left|\binom{\mu}{k}\right| \leq \mu \quad \text{for} \quad k \geq 1,$$

$$\text{so} \quad \|E\| \leq \sum_{k=1}^{\infty} \mu \, \|C\|^k \leq \mu\|C\|(1-\epsilon)^{-1} \quad \text{and} \tag{3.52}$$

$$\|F\| \leq \mu \, \|D\|(1-\epsilon)^{-1}. \tag{3.53}$$

It follows that if $\delta > 0$ is as in Lemma 3.4, there exists $\epsilon > 0$, independent of $\lambda \geq 1$, such that if $\|(C,D)\| < \epsilon$ then $\|(E,F)\| < \delta$. Furthermore, by taking δ sufficiently small we can assume that

$$B^{\mu}(B^{-\mu}A^{\mu})^{\frac{1}{2}} = I + W \quad \text{with} \quad \|W\| < 1 \tag{3.54}$$

It follows that for $(A,B) \in U$ we can write $f_{\lambda}(A,B)$ as the composition of the analytic maps $(A,B) \to (A^{\mu},B^{\mu}) = (A_1,B_1)$, $(A_1,B_1) \to A_2 = B_1(B_1^{-1}A_1)^{\frac{1}{2}}$ and $A_2 \to A_2^{\lambda}$, so f_{λ} is analytic on U.

If $\eta = \sup(\|C\|,\|D\|)$ and W is as in (3.54), we obtain from (3.52) and (3.53) and Lemma 3.4 that there exists a constant M_1, independent of $\lambda \geq 1$, such that

$$\|W\| \leq \mu\eta M_1. \tag{3.55}$$

The binomial theorem gives

$$(I+W)^{\lambda} = I + \sum_{k=1}^{3} \binom{\lambda}{k} W^k + \sum_{k=4}^{\infty} \binom{\lambda}{k} W^k \tag{3.56}$$

However we have

$$\left\| \sum_{k=4}^{\infty} \binom{\lambda}{k} W^k \right\| \leq \sum_{k=4}^{\infty} \left| \binom{\lambda}{k} \right| \mu^k \eta^k M_1^k \tag{3.57}$$

It is an easy exercise to see that

$$\left| \binom{\lambda}{k} \mu^k \right| \leq 1,$$

so we find (possibly after decreasing ϵ) that

$$\left\| \sum_{k=4}^{\infty} \binom{\lambda}{k} W^k \right\| \leq \sum_{k=4}^{\infty} \eta^k M_1^k \leq \frac{\eta^4 M_1^4}{1-\epsilon M_1} \tag{3.58}$$

It follows from equations (3.56) that to prove Theorem 3.2 it suffices to prove that $I + \sum_{k=1}^{3} \binom{\lambda}{k} W^k$ equals the right hand side of (3.48), but with a different remainder term $R_5(\mu,C,D)$ which also satisfies (3.49).

Recall that E and F are given by equations (3.50) and (3.51) and W is then given by (3.38). If we recall equations (3.52) and (3.53) and use the standard "big oh" notation (with the understanding that constants in the "big oh" notation are independent of $\lambda \geq 1$) we find

$$I + \sum_{k=1}^{3} \binom{\lambda}{k} W^k = I + \left(\frac{\lambda}{2}\right)(E+F) - \left(\frac{\lambda}{8}\right)(E-F)^2 + \left(\frac{\lambda}{16}\right)(E-F)(E+F)(E-F)$$

$$+ \frac{\lambda(\lambda-1)}{8}(E+F)^2 - \frac{\lambda(\lambda-1)}{32}(E+F)(E-F)^2 \tag{3.59}$$

$$- \frac{\lambda(\lambda-1)}{32}(E-F)^2(E+F) + \frac{\lambda(\lambda-1)(\lambda-2)}{48}(E+F)^3 + 0(\eta^4)$$

By using equations (3.50) and (3.51) one can express each of the terms on the right hand side of (3.59) by an expression involving powers of C and D and a term $0(\eta^4)$. For example, one finds that

$$(E+F)^2 = \mu^2(C+D)^2 + (\tfrac{1}{2})\mu^2(\mu-1)(C+D)(C^2+D^2)$$

$$+ (\tfrac{1}{2})\mu^2(\mu-1)(C^2+D^2)(C+D) + 0(\eta^4) \qquad (3.60)$$

If one combines equations (3.55) – (3.59) and uses expression like equation (3.60) for the terms in eq. (3.59), then, after a lengthy and unpleasant calculation, one obtains eq. (3.48). □

If $C,D \in L(H)$ and $\|(C,D)\| \leq r < 1$, one can prove that there exists $\lambda(r) \geq 1$ such that (setting $A = I + C$, $B = I + D$ and $\mu = \lambda^{-1}$),

$$(B^{\mu}(B^{-\mu}A^{\mu})^{\frac{1}{2}})^{\lambda}$$

is defined for all $\lambda \geq \lambda(r)$. One sees this by essentially the argument used in Theorem 4.3 of [28]. Furthermore, the argument in Theorem 4.3 also shows that for $r < 1$ and $\|(C,D)\| \leq r$

$$n - \lim_{\lambda\to\infty}(B^{\mu}(B^{-\mu}A^{\mu})^{\frac{1}{2}})^{\lambda} = \exp(\tfrac{1}{2}\log A + \tfrac{1}{2}\log B)$$

and that the convergence is uniform in (C,D) such that $\|(C,D)\| \leq r$. It follows that one also has convergence of the Taylor series as $\lambda \to \infty$, and eq. (3.48) remains valid for $\mu = 0$ on U if one replaces the left–hand side of eq. (3.48) by $\exp(\frac{1}{2}\log A + \frac{1}{2}\log B)$.

As an immediate consequence of Theorem 3.2 we can prove that inequality (3.26) usually holds.

COROLLARY 3.1 Let assumptions and notation be as in Theorem 3.2. For $C,D \in \mathscr{L}(H)$ such that $\|(C,D)\| < \epsilon$ (ϵ as in Theorem 3.2) and for complex numbers z with $|z| \leq 1$, define

$$A_z = I + zC \quad \text{and} \quad B_z = I + zD.$$

For C and D as above and $\lambda \geq 1$ and $|z| \leq 1$, define $\mu = \lambda^{-1}$ and

$$[B_z^{\mu}(B_z^{-\mu}A_z^{\mu})^{\frac{1}{2}}]^{\lambda} = \varphi(\lambda,z) \quad \text{and}$$

$$\exp(\tfrac{1}{2}\log(A_z) + \tfrac{1}{2}\log(B_z)) = \varphi(\infty,z)$$

Define E by

$$E = [[C,D],C-D] = \{[C,D]C + C[D,C] + D[C,D] + [D,C]D\}.$$

If $E \neq 0$ and $1 \leq \lambda_1 < \lambda_2 \leq \infty$, there are at most finitely many values of z with $|z| \leq 1$ such that $\varphi(\lambda_1,z) = \varphi(\lambda_2,z)$.

Proof. If $E \neq 0$, there exist x and $y \in H$ such that

$$\langle Ex,y\rangle \neq 0.$$

Furthermore, for any $\lambda \geq 1$ the map $z \to \langle\varphi(\lambda,z)x,y\rangle = \psi(\lambda,z)$ is an analytic map defined on an open neighborhood of the closed unit disc in $\mathbb{C}$. Theorem 3.2 shows that if $1 \leq \lambda_1 < \lambda_2$, the cubic term of the Taylor polynomial for $\psi(\lambda,z)$ centered at $z = 0$ is unequal to the cubic term of the Taylor polynomial for $\psi(\lambda_2,z)$. Because of analyticity, it follows that $\psi(\lambda_1,z)$ equals $\psi(\lambda_2,z)$ for at most finitely many values of z in the closed unit disc. □

Specializing to the case $\lambda_1 = 1$ and $\lambda_2 = \infty$, Corollary 3.1 provides an answer to a question raised by G.E. Trapp [30, pp. 119–120].

We now return to the problem of proving that the operators $E_\lambda = E_\lambda(A,B)$ in eq. (3.14) are, in general, unequal for unequal values of $\lambda \geq 1$. Before stating our precise theorem, it is convenient to recall a well-known fact about analytic functions which is also true in the Banach space case. Suppose that Y is a complex Banach space and $h: B_r(y^0) \to Y$ is analytic and $\|h(y)\| \leq M$ for all $y \in B_r(y^0)$. Suppose that $T_k(y)$ denotes the Taylor polynomial for $h(y)$ of degree k, so

$$h(y) = T_k(y) + R_k(y)$$

Then for $\|y\| \leq r_1 < r$ and M as above one has

$$\|R_k(y)\| \leq \frac{M\|y\|^{k+1}}{(r-r_1)r^k}.$$

THEOREM 3.3 Let H be a complex Hilbert space, $\lambda \geq 1$ a real number and $\mu = \lambda^{-1}$. For $\rho > 0$ define $U_\rho = \{(A,B)\colon A,B \in \mathscr{L}(H), \|A-I\| < \rho$ and $\|B-I\| < \rho\}$. For $\lambda \geq 1$ let $f_\lambda(A,B)$ be defined as in the statement of Theorem 3.2 and define f_∞ by

$$f_\infty(A,B) = \left(\frac{A+B}{2}, \exp(\tfrac{1}{2}\log A + \tfrac{1}{2}\log B)\right)$$

By Theorem 3.2 there exist $\epsilon > 0$ so that f_λ is defined and analytic on U_ϵ for all $\lambda \geq 1$. There exists $r > 0$, independent of $\lambda \geq 1$, such that for all $\lambda \geq 1$, f_λ^k, the kth iterate of f_λ, is defined on $\overline{U}_r$, $f_\lambda^k(\overline{U}_r) \subset U_\epsilon$ for all $k \geq 1$, and

$$n - \lim_{k\to\infty} f_\lambda^k(A,B) = (E_\lambda, E_\lambda), \tag{3.61}$$

where $E_\lambda = E_\lambda(A,B)$ is an analytic function of $(A,B) \in U_r$. The convergence in eq. (3.61) is uniform for $(A,B) \in \overline{U}_r$. Furthermore, if

$A = I + C$ and $B = I + D$ and $\|C\| \le r_1 < r$ and $\|D\| \le r_1 < r$, one has that

$$E_\lambda(A,B) = I + (\frac{C+D}{2}) - (\frac{1}{16})(C-D)^2 + (\frac{1}{32})(C-D)(C+D)(C-D)$$

$$((\frac{\mu-1}{48}) + \frac{(\mu-1)^2}{96}) \, [[C,D],C-D] + R(\mu,C,D), \tag{3.62}$$

and the remainder term $R(\mu,C,D)$ satisfies

$$\|R(\mu,C,D)\| \le \frac{(1+\epsilon)\|(C,D)\|^4}{r^3(r-r_1)} \tag{3.63}$$

Proof. If $\|(C,D)\| < \epsilon$, Theorem 3.2 implies that for fixed $\lambda \ge 1$

$$f_\lambda(A,B) = g(A,B) + R_1(A,B), \text{ where}$$

$$g(A,B) = (\frac{A+B}{2}, g_2(A,B)) \quad \text{and}$$

$$g_2(A,B) = \frac{A+B}{2} - (\frac{1}{8})(C-D)^2 + (\frac{1}{16})(C-D)(C+D)(C-D) \tag{3.64}$$

$$+ (\frac{1}{24})((\mu-1) + (\frac{1}{2})(\mu-1)^2)[[C,D],C-D]$$

The remainder term $R_1(A,B)$ satisfies

$$R_1(A,B) = (0, \psi_1(A,B)), \text{ where}$$

$$\|\psi_1(A,B)\| \le M(\|C\|^2+\|D\|^2)^2 \quad \text{and} \tag{3.65}$$

the constant M is independent of $\lambda \ge 1$.

The idea of the proof is to use Lemma 3.3. If $S = \{(A,A): A \in \mathscr{L}(H)\}$, we clearly have that $f_\lambda(A,A) = g(A,A) = (A,A)$ for

$(A,A) \in S \cap U_\epsilon$, so that

$$R_1(A,B) = 0 \quad \text{for} \quad (A,B) \in S \cap U_\epsilon.$$

We know R_1 is analytic, and inequality (3.65) implies that the Fréchet derivative of R_1 at (I,I) equals zero and that $\|R_1(A,B)\|$ is bounded on U_ϵ by a constant independent of $\lambda \geq 1$.

It remains to show that g satisfies the conditions of Lemma 3.2. If $y = (A,B)$ and $p(y) = \|A-B\|$, we have from (3.64) that

$$p(g(y)) \leq (\tfrac{1}{8})\|C-D\|^2 + (\tfrac{1}{16})\|C-D\|^2\|C+D\| \tag{3.66}$$

$$+ (\tfrac{1}{24})\|CD-DC\| \ \|C-D\|$$

Since we have

$$[C,D] = (\tfrac{1}{2})[C-D,\ C+D], \tag{3.67}$$

we conclude from (3.66) and (3.67) that

$$p(g(y)) \leq (\tfrac{1}{8})p(y)^2 + (\tfrac{1}{16})p(y)^2\|C+D\| + (\tfrac{1}{24})p(y)^2\|C+D\|. \tag{3.68}$$

If $\epsilon \leq 1$, $p(y) \leq 2$ and $\|C+D\| \leq 2$ and (3.68) yields,

$$p(g(y)) \leq (\tfrac{2}{3})p(y). \tag{3.69}$$

We still must estimate $\|g(y)-y\|$. We have

$$\|(\tfrac{A+B}{2}) - A\| = \|\tfrac{A+B}{2} - B\| = (\tfrac{1}{2})p(y). \tag{3.70}$$

Also, using eq. (3.69) and (3.70) we see that

$$\|g_2(A,B) - B\| = \|g_2(A,B) - (\frac{A+B}{2}) + (\frac{A+B}{2}) - B\|$$

$$\leq p(g(y)) + \frac{1}{2}p(y) \leq (\frac{2}{3} + \frac{1}{2})\ p(y). \tag{3.71}$$

We conclude from (3.70) and (3.71) that

$$\|g(y) - y\| \leq kp(y),\ k = \frac{7}{6}. \tag{3.72}$$

We have now verified the hypotheses of Lemma 3.3, so there exists a number $r > 0$ as in the statement of Theorem 3.3 and eq. (3.61) is satisfied. According to Lemma 3.3, r depends on constants which we have shown can be chosen independent of $\lambda \geq 1$, so r is independent of $\lambda \geq 1$. We know that $E_\lambda(A,B)$ is an analytic function of $(A,B) \in U_r$ and that $\|E_\lambda(A,B)\| \leq (1+\epsilon)$ for $(A,B) \in U_r$, so the estimate in (3.63) will follow if we can prove that the terms in eq. (3.62) give the Taylor polynomial of degree 3 for $E_\lambda(A,B)$. By using eq. (3.61) and analyticity, one can prove that the Taylor polynomial of degree three for $f_\lambda^k(A,B)$ converges in norm to the Taylor polynomial of degree three for $E_\lambda(A,B)$.

Define G by

$$G = (\frac{C+D}{2}) - (\frac{1}{16})(C-D)^2 + (\frac{1}{32})(C-D)(C+D)(C-D)$$

$$+ (\frac{1}{48})((\mu-1) + \frac{1}{2}(\mu-1)^2)[[C,D],\ C-D].$$

Fix $r_1 < r$, $r_1 > 0$, and suppose we can prove that for $\|C\| \leq r_1$ and $\|D\| \leq r_1$ and $k \geq 2$ we have

$$f_\lambda^k(A,B) = (I+G+\theta_k(A,B),\ I+G+\psi_k(A,B)), \tag{3.73}$$

where $\|\theta_k(A,B)\| \leq M_k(\|C\|+\|D\|)^4$ and

$$\|\psi_k(A,B)\| \le M_k(\|C\|+\|D\|)^4 \tag{3.74}$$

Then it will follow that $I + G$ is the Taylor polynomial of degree three for f_λ^k, and by the remarks above the theorem will follow.

If $A = I+C$ and $B = I+D$ with $\|(C,D)\| < r$, we define

$$f_\lambda^k(A,B) = (I+C_k,\ I+D_k) = (A_k,B_k)$$

The uniform boundedness of f_λ^k on U_r for $\lambda \ge 1$ and $k \ge 1$ and the analyticity of f_λ^k implies a uniform bound on the norm of the Fréchet derivative of f_λ^k on U_{r_1} ($r_1 < r$). Using this bound we find that there exists a constant L, independent of $\lambda \ge 1$ and $k \ge 1$, such that for $\|(C,D)\| \le r_1$ we have

$$\|(C_k,D_k)\| \le L\|(C,D)\| \tag{3.75}$$

Notice also that eq. (3.68) implies that

$$\|C_1-D_1\| \le (\tfrac{1}{8} + \tfrac{r}{8} + \tfrac{r}{12})\|C-D\|^2 \tag{3.76}$$

If we use eq. (3.64) and substitute C_1 for C and D_1 for D we find that

$$C_2 = G + (\frac{\psi_1(A,B)}{2}) \quad \text{and} \tag{3.77}$$

$$D_2 = G + \psi_2(A,B), \quad \text{where} \tag{3.78}$$

$$\psi_2(A,B) = -\tfrac{1}{8}(C_1-D_1)^2 + \tfrac{1}{16}(C_1-D_1)(C_1+D_1)(C_1-D_1)$$

$$+ \tfrac{1}{48}((\mu-1) + \tfrac{1}{2}(\mu-1)^2)\ [[C_1-D_1,\ C_1+D_1],\ C_1-D_1]$$

$$+ \frac{\psi_1(A,B)}{2} + \psi_1(A_1,B_1)$$

By using eqns. (3.65), (3.75) and (3.76) we see that there exist a constant M_2 such that

$$\|\psi_2(A,B)\| \leq M_2(\|C\| + \|D\|)^4.$$

Assume that equations (3.73) and (3.74) are satisfied for some $k \geq 1$ and all (C,D) with $\|(C,D)\| \leq r_1$. By using eq. (3.64) (with C_k replacing C and D_k replacing D) we obtain

$$C_{k+1} = G+\theta_{k+1}(A,B),\ \theta_{k+1}(A,B) = (\tfrac{1}{2})\theta_k(A,B)+(\tfrac{1}{2})\psi_k(A,B) \text{ and} \tag{3.80}$$

$$D_{k+1} = G + \psi_{k+1}(A,B), \text{ where} \tag{3.81}$$

$$\psi_{k+1}(A,B) = -\tfrac{1}{8}(\theta_k-\psi_k)^2 + \tfrac{1}{16}(\theta_k-\psi_k)(2G+\psi_k+\theta_k)(\theta_k-\psi_k) \tag{3.82}$$

$$+ \tfrac{1}{48}((\mu-1) + (\tfrac{1}{2})(\mu-1)^2)[[\theta_k-\psi_k,\ 2G + \theta_k + \psi_k],\ \theta_k - \psi_k]$$

$$+ \theta_{k+1}(A,B) + \psi_1(A_k,B_k)$$

Again, with the aid of eqns. (3.65), (3.75) and (3.76) we see that there exists a constant M_{k+1} such that

$$\|\theta_{k+1}(A,B)\| \leq M_{k+1}(\|C\|+\|D\|)^4 \quad\text{and}\quad \|\psi_{k+1}(A,B)\| \leq M_{k+1}(\|C\|+\|D\|)^4.$$

By induction, equations (3.73) and (3.74) are satisfied for all $k \geq 2$, and the proof is complete. □

Notice that we have actually proved in Theorem 3.3 that there exists a constant M_1 (independent of $\lambda \geq 1$) such that for $y \in U_\epsilon$ we have $p(f_\lambda(y)) \leq M_1(p(y))^2$. Under these circumstances it is proved in Lemma 3.2 of [28] that convergence of $f_\lambda^k(y)$ as $k \to \infty$ is quadratic and hence extremely rapid.

Just as in Corollary 3.1, we can use Theorem 3.3 to prove that in general the operators $E_\lambda(A,B)$ are distinct for distinct values of $\lambda \geq 1$.

COROLLARY 3.2 Let assumptions and notations be as in Theorem 3.3. For r as in the statement of Theorem 3.3 assume that $C, D \in \mathscr{L}(H)$ and that $\|C\| < r$ and $\|D\| < r$. For complex numbers z such that $|z| \leq 1$ define $A_z = I + zC$ and $B_z = I + zD$ and define $E_\lambda(A_z,B_z)$ for $\lambda \geq 1$ by

$$E_\lambda(A_z,B_z) = n - \lim_{k\to\infty} f_\lambda^k(A_z,B_z).$$

Assume that

$$[[C,D],\ C-D] \neq 0.$$

Then if $1 \leq \lambda_1 < \lambda_2 \leq \infty$, we have

$$E_{\lambda_1}(A_z,B_z) \neq E_{\lambda_2}(A_z,B_z)$$

except for at most finitely many numbers z with $|z| \leq 1$.

Proof. With the aid of Theorem 3.3, Corollary 3.3 follows by essentially the same argument used to prove Corollary 3.1. Details are left to the reader. □

REFERENCES

1. W.N. Anderson, Jr. and R.J. Duffin, Series and parallel addition of matrices, J. Math. Anal. Appl. 26(1969), 576–594.

2. W.N. Anderson, Jr., and G.E. Trapp, Shorted operators II, SIAM J. Appl. Math. 28 (1975), 60–71.

3. W.N. Anderson, Jr., and G.E. Trapp, Operator means and electrical networks, Proc. 1980 IEEE International Symposium on Circuits and Systems (1980), 523–527.

4. Tsuyoski Ando, Topics on Operator Inequalities, Lecture Notes, Hokkaido University, Sapporo, 1978.

5. ______________, Concavity of certain maps on positive definite matrices and applications to Hadamard products, Linear Algebra and its Applications 27(1979), 203–241.

6. ______________, On the arithmetic–geometric–harmonic mean inequality for positive definite metrices, Linear Algebra and its Applications 52–53 (1983), 31–37.

7. J. Arazy, T. Claesson, S. Janson and J. Peetre, Means and their iterations, in Proceedings of the Nineteenth Nordic Congress of Mathemticians, Reykjavik, 1984, 191–212 (published by the Icelandic Math. Soc., 1985).

8. Richard Beals, Topics in Operator Theory, University of Chicago Press, Chicago, Illinois, 1971.

9. J. Bendat and S. Sherman, Monotone and convex operator functions, Trans. Amer. Math. Soc. 79 (1955), 58–71.

10. K.V. Bhagwat and R. Subramanian, Inequalities between means of positive operators, Proc. Cambridge Phil. Soc. 83(1978), 393–401.

11. G. Borchardt, Gesammelte Werke, herausgegeben von G. Hettner, Reimer, Berlin, 1888.

12. H.J. Carlin and G.A. Noble, Circuit properties of coupled dispersive lines with applications to wave guide modeling, Proc. Network and Signal Theory, edited by J.K. Skwirzynski and J.O. Scanlan, Peter Pergrinus, Inc., London, 1973, 258–269.

13. B.C. Carlson, Algorithms involving arithmetic and geometric means, Amer. Math. Monthly 78(1971), 496–504.

14. J.E. Cohen and R.D. Nussbaum, Arithmetic–geometric means of positive matrices, Math. Proc. Cambridge Phil. Soc., 101(1987), 209–219.

15. D.A. Cox, The arithmetic–geometric mean of Gauss, Enseignement Math. 30(1984), 275–330.

16. J. Dieudonne, Foundations of Modern Analysis, Academic Press, New York, 1960.

17. W. Donoghue, Monotone Matrix Functions and Analytic Continuation, Springer–Verlag, New York, 1974.

18. R.G. Douglas, On majorization, factorization and range inclusion of operators on Hilbert space, Proc. A.M.S. 17(1966), 413–415.

19. Jun–ichi Fujii, Arithmetico–geometric mean of operators, Math. Japonica 23(1979), 667–669.

20. ____________________, On geometric and harmonic means of positive operators, Math. Japonica 24(1979), 203–207.

21. G.H. Hardy, J.E. Littlewood and G. Polya, Inequalities, 2nd ed., Cambridge University Press, 1959.

22. F.Kubo and T. Ando, Means of positive linear operators, Math. Annalen 246(1980), 205–224.

23. K.Löwner=C.Loewner, Uber monotone Matrixfunktionen, Math. Zeit. 38(1934), 177–216.

24. C. Loewner, Some classes of functions defined by difference or differential inequalities, Bull. Amer. Math. Soc. 56(1950), 308–319.

25. R.D. Nussbaum, Iterated nonlinear maps and Hilbert's projective metric, to appear in Memoirs of the AMS.

26. ____________________, "Iterated nonlinear maps and Hilbert's projective metric, II", submitted for publication.

27. ____________________, "Iterated nonlinear maps and Hilbert's projective metric: a summary",in Dynamics of Infinite Dimensional Systems, edited by Shui-nee Chow and Jack K. Hale, NATO ASI Series, Springer-Verlag, Heidelberg, 1987, pp. 231-249.

28. R.D. Nussbaum and J.E. Cohen, The arthmetic-geometric mean and its generalizations for noncommuting linear operators, to appear in Annali della Scuola Normale Sup. di. Pisa.

29. W. Pusz and S.L. Woronowicz, Functional calculus for sesquilinear forms, and the purification map, Rep. Math. Phys. 8(1975), 159-170.

30. G.E. Trapp, Hermitian semidefinite matrix means and related matrix inequalities - an introduction, Linear and Multilinear Algebra 16(1984), 113-123.

31. J. Wimp, Computation with Recurrence Relations, Pittman Advanced Publishing Program, Boston, 1984.

32. K. Yosida, Functional Analysis, sixth ed., Springer-Verlag, New York, 1980.

Recent Progress in Periodic Orbits of Autonomous Hamiltonian Systems and Applications to Symplectic Geometry

CLAUDE VITERBO Ceremade, UA 749, Université de Paris 9,
Place du Marechal de Lattre de Tassigny, F-75775 PARIS, FRANCE
and
Visiting Member, MSRI, 1000 Centennial Drive, Berkeley CA 94720, USA.

Let (P^{2n}, ω) be a symplectic manifold, and $H : P \to \mathbb{R}$ be a smooth function, the Hamiltonian vector field X_H of H is defined by

$$\omega(X_H(x), \xi) = dH(x)\xi \text{ for all } \xi \in T_x P.$$

The main properties of the flow of X_H are that it preserves both the form ω and the "energy levels" $\{x \mid H(x) = c\}$. The system $\dot{x} = X_H(x)$ is called an autonomous Hamiltonian system.

The aim of this survey paper is to introduce the reader to recent results and techniques in periodic orbits for autonomous Hamiltonian systems and their applications to symplectic geometry.

In order to deal with one difficulty at a time, we shall only work in the symplectic manifold $(\mathbb{R}^{2n}, \sigma)$ where $\sigma = \sum_{i=1}^{n} dx_i \wedge dy^i$, except in section 4 where we need to work in more general manifolds. This allows us to introduce Floer homology (cf. [**Fl 1**], [**Fl 2**]) very simply (of course on a general manifold one needs the more complicated approach of [**Fl 1**], [**Fl 2**]) through the Amman–Conley– Zehnder finite dimensional reduction.

We then give a proof of the existence of a periodic orbit on any compact hypersurface of contact type (Weinstein's conjecture (cf. [**V 2**])).

We also apply these methods to define the Ekeland-Hofer capacities and mention some of their applications to symplectic geometry. We then explain how to apply our method to show nonexistence of exact lagrangian torii, first proved by Gromov [**G 1**], using holomorphic curves.

We conclude by explaining a different method to prove the Weinstein's conjecture in compact manifolds, its applications, and finally some conjectures.

A word is necessary to warn the reader about the content of this paper. The author has mainly emphasized his own work and related results; we obviously do not pretend here to cover all aspects of autonomous Hamiltonian systems or symplectic geometry !

I would like to thank for helpful discussions and comments Y. Eliashberg, A.Uribe and A. Weinstein. Special thanks to Dusa Mac Duff for reading the manuscript, and many useful suggestions.

1 HISTORICAL AND MATHEMATICAL MOTIVATIONS

Interest in periodic orbits for Hamiltonian systems goes back at least to Poincaré. Poincaré was the first to give convincing evidence (prove?) that a general Hamiltonian system could not be integrated, the question then arose of how to describe the Hamiltonian flow. So far, it seems that the only relevant properties of the flow are

- the ergodic properties of the flow
- the existence of invariant torii (KAM theory)
- the existence of periodic orbits

Another reason to look for periodic orbits is the relation between pseudo - differential operators on a manifold M and Hamiltonian systems on $(T^*M, dp\wedge dq)$. To each operator we associate its principal symbol that we consider as a Hamiltonian on T^*M(e.g. to the Laplace operator corresponding to the metric g, we associate $H(q,p) = g(q)(p,p)$).

Now assuming the operator A to be self-adjoint and positive elliptic, let $\lambda_1 \leq \lambda_2 \leq \lambda_3 \leq \dots$ be its eigenvalues; we consider the distribution

$$\varphi(t) = \sum_{k=1}^{+\infty} e^{i(\lambda_k)^{1/m}t} \quad \text{“} = \operatorname{Tr} e^{it(A)^{1/m}} \text{”}$$

where m is the order of A.

Then according to Chazarain, Duistermaat and Guillemin([**Ch**] [**D-G**]), the singularities of φ are either 0, or periods of closed trajectories of the associated Hamiltonian on the energy level $\{(q,p) \mid H(q,p) = 1\}$.

Even though it is not completely clear what knowledge of the singularities of φ actually says about A, it is nonetheless clear that periodic orbits are important objects.

Finally we shall see in section 6 how the problem of finding periodic orbits can be considered as a special case of a conjecture on isotropic foliations whose other special case is the nonexistence of exact lagrangian embeddings.

We now turn to the methods used to find periodic orbits. Until Rabinowitz's paper [**R**], the only methods were either perturbation methods or special geometric methods (cf. [**Se**], [**We 1**]) which enabled A. Weinstein to prove existence of at least one periodic orbit on any convex energy hypersurface.

From our point of view, the modern approach started with Rabinowitz's paper in which he considers periodic solutions of $\dot{x} = X_H(x)$ as critical points of

$$A_H(x) = \int_0^T \left[\frac{1}{2}(J\dot{x}, x) + H(x)\right] dt$$

on the space $H^{1/2}(S^1, \mathbb{R}^{2n})$.

Even though this has been long known as Maupertuis's principle, the fact that any critical point of A_H has infinite Morse index and coindex, so that the topology of $\{x \in H^{1/2}(S^1, \mathbb{R}^{2n}) \mid A_H(x) \leq c\}$ does not depend on c, and thus makes critical levels topologically undetectable had deterred many people before. As we will show in section 2, there is in fact a change in topology, but in a more subtle way.

The methods used to deal with A_H quickly improved, from Rabinowitz's Galerkin approximation method followed by Clarke and Ekeland's dual action functional ([**C-E**]) until Amman–Zehnder Lyapunov Schmidt reduction ([**A-Z 2**]), Chaperon's broken geodesics method ([**C**]) and finally the holomorphic curve approach of Gromov and Floer ([**G 1**], [**G 2**], [**Fl 1**], [**Fl 2**]) which allows one to work on general symplectic manifolds.

In the next section, we shall essentially present the Amman– Zehnder method which appears to us as the best suited for working in $(\mathbb{R}^{2n}, \sigma)$.

2 THE AMANN- ZEHNDER REDUCTION

The aim of this section is to replace A_H by a function a_N defined on a finite dimensional vector space E_N (of dimension $2nN$) such that there is a one to one correspondence between critical points of A_H and critical points of a_N. We shall also see that the critical levels do not depend on N, and that the Morse indices and the topology behave "nicely" as N changes (cf. Proposition 2.1, 2.2).

Throughout this section we make the assumption that H is smooth and $|H''(x)| \leq C$ on $\mathbb{R}^{2n}$.

Let us first define a_N.

For $x \in E = H^1(S^1, \mathbb{R}^{2n}) = \{x \in L^2(S^1, \mathbb{R}^{2n}) \mid \dot{x} \in L^2\}$, consider its Fourier series decomposition

$$x = \sum_{k \in \mathbb{Z}} \exp(kJt) x_k \qquad x_k \in \mathbb{R}^{2n}$$

and set $x = u_N + v_N$ where

$$u_N = \sum_{k=-N}^{N} \exp(kJt)x_k$$

$$v_N = \sum_{|k|>N} \exp(kJt)x_k$$

and define the projections P_N, Q_N, by $P_N x = u_N$, $Q_N u = v_N$. Finally set $E_N = \operatorname{Im} P_N = \ker Q_N$, $F_N = \operatorname{Im} Q_N = \ker P_N$, and D be the self-adjoint unbounded operator $Dx = J\dot{x}$.

Having fixed the notations, we state

PROPOSITION 2.1.

(1) *There is a* C^1 *map* $w_N : E_N \to F_N$ *such that* $v_N = w_N(u)$ *is the unique solution of* $\frac{\partial}{\partial v} A_H(u_N + v_N) = 0$

(2) *If we set* $a_N(u) = A_H(u + w_N(u))$ *then* a_N *is a* C^2 *function on* E_N *such that*

- (i) u *is a critical point of* a_N *if and only if there is a* v *in* F_N *such that* $u + v$ *is a critical point of* A_H
- (ii) *such a* v *is unique and equal to* $w_N(u)$

(3) *The canonical* S^1 *action on* E *defined by* $\theta \cdot x(t) = x(t + \theta)$ *restricts to an* S^1 *action on* E_N *which makes* w_N *and thus* a_N *equivariant.*

PROOF:

(1) We write down $\nabla A_H(u + v)$ as

$$J\dot{u} + J\dot{v} + P_N \nabla H(u + v) + Q_N \nabla H(u + v)$$

so that

$$\begin{aligned} \nabla_v A_H(u + v) &= J\dot{v} + Q_N \nabla H(u + v) \\ &= Dv + Q_N \nabla H(u + v). \end{aligned}$$

Now since the Fourier coefficients of v with index in $[-N, N]$ vanish, we have that $\|Dv\|_{L^2} \geq N\|v\|_{L^2}$, so that $\|D^{-1}\| \leq \frac{1}{N}$. Moreover, $v \to \nabla H(u + v)$ is Lipschitz with Lipschitz norm less than C, so that $v \to -D^{-1} Q_N \nabla H(u +$

v) is a continuous family of Lipschitz maps with Lipschitz norm less than $\frac{1}{N} \cdot C$. From now on we assume $C < \frac{N}{2}$, so that by the shrinking lemma $v = -D^{-1}Q_N \nabla H(u+v)$ determines a unique $v = w_N(u)$ which is at least continuous in u(and in fact one easily proves that w_N is C^1).

(2) We have

$$\begin{aligned} Da_N(u) &= D(A_H(u + w_N(u))) \\ &= \frac{\partial}{\partial u} A_H(u + w_N(u)) \\ &+ \frac{\partial}{\partial v} A_H(u + w_N(u)) w'_N(u) \end{aligned}$$

since the second term on the right hand side is zero, we infer that $Da_N(u) = \frac{\partial}{\partial u} A_H(u + w_N(u))$. From this, (i) and (ii) easily follows.

(3) is quite obvious, since we did not break the S^1-symmetry anywhere.

□

Of course, according to the above proposition, in order to find a critical orbit of A_H, we only have to find a critical orbit of a_N for some given N. However, if we really want to understand A_H, we must understand how the a_N relate to each other for different values of N.

Our tool to find critical orbits of a_N will be the following. First, on a space X on which a function f is implicitly defined, we set $X^a = \{x \in X \mid f(x) \leq a\}$. Here we moreover assume that X carries an S^1 action.

PROPOSITION 2.2 ([**V** 4]). *Let f be an S^1 invariant function on some space X. If $H^d_{S^1}(X^b, X^a) \neq 0$, then f has at least one critical value in $[a, b]$. Moreover, if the critical orbits of f are isolated, and there is no fixed point of the action in $X^b - X^a$, then there is a critical level c in $[a, b]$ such that for some critical orbit in $f^{-1}(c)$, of Morse index m and nullity n, we have*

$$m \leq d \leq m + n.$$

The proof is quite obvious, and expresses the fact that a change in the "d-dimensional topology" of the X^λ can only be due to a critical orbit with appropriate index.

In view of this, the following two results are quite naturally related.

PROPOSITION 2.3. *Assume $N' \geq N > 2C$, then we have*

$$H_{S^1}^{*-2nN}(E_N^b, E_N^a) \approx H_{S^1}^{*-2nN'}(E_{N'}^b, E_{N'}^a)$$

for any pair $a < b$ of regular values of a_N.

Here $H_{S^1}^*(X)$ is the Borel equivariant cohomology $H^*(X \times S^\infty/S^1)$ where the S^1 action is the diagonal action on $X \times S^\infty$ and S^∞ is the unit sphere in $L^2(S^1)$ with the canonical S^1 action.

The proof is given in [**V 5**]. The above isomorphism is "induced" by the map from E_N^λ into $E_{N'}^\lambda$ given by $j_N : u_N \to u_N + P_{N'}(w_N(u_N))$. We then prove that we can find a pseudogradient vector field for $a_{N'}$ which coincides with ∇a_N on the image of j_N, and "well behaved" outside a neighborhood of this image.

Finally we use Conley's Morse index theory ([**Co**]) to compare the invariant sets
$E_{N'}^b - E_{N'}^a$ and $j_N(E_N^b - E_N^a)$, and see that the first is the suspension of the second.

The companion result to Proposition 2.3 is, in view of 2.2, the following

PROPOSITION 2.4. *Let u be a critical orbit of a_N and $u' = u + P_{N'}(w_N(u))$ the corresponding critical orbit of $a_{N'}$. Then if $i_N, i_{N'}$ and $\nu_N, \nu_{N'}$ are the Morse coindices and nullities of u and u', we have*

(i) $\nu_N = \nu_{N'}$
(ii) $i_{N'} - 2nN' = i_N - 2nN$

The proof is quite simple and can be essentially found in [**Co-Z**](see also [**V 1**]). Our choice of the coindex instead of the index is meant to agree with the Conley-Zehnder and Ekeland definitions of index. Now, in view of 2.3, 2.4 it is natural to set

DEFINITION 2.5.

$$FH_{S^1}^*(E^b, E^a) = H_{S^1}^{*-2nN}(E_N^b, E_N^a)$$

$$i(x) = i_N(P_N(x)) - 2nN$$

$$\nu(x) = \nu_N(P_N(x))$$

where x is a critical orbit of A_H.

REMARK: The notation FH^* stands for Floer homology.

This was defined in [**Fl 1**], [**Fl 2**] in the case of a time dependent Hamiltonian with nondegenerate fixed points (so that we have no S^1 action) as the cohomology of the following complex (C^*, δ^*)

$$C^n = \text{ free group with generators the critical points of } A_H \text{ with } i(x) = n$$

$$\delta^n x = \sum_{y \in C^{n+1}} \langle \delta^n x, y \rangle y$$

where $\langle \delta^n x, y \rangle$ is the algebraic number of solutions of the elliptic system

$$(2.6) \qquad \begin{cases} \left(\frac{\partial}{\partial t} + J\frac{\partial}{\partial \theta}\right) u = -\nabla H(t, u) \\ u : \mathbb{R} \times S^1 \to P \\ \lim_{t \to -\infty} u(t, \theta) = x(\theta) \quad \lim_{t \to +\infty} u(t, \theta) = y(\theta) \end{cases}$$

where J is an almost complex structure on the symplectic manifold (P, ω).

Note that equation (2.6) is just the equation of the "L^2 gradient flow of A_H" and if E were a finite dimensional manifold it is well known that the cohomology of (C^*, δ^*) called the Morse Smale complex of A_H is the cohomology of E (cf. [**M**]). In our case we can replace (2.6) by the flow equation of ∇a_N, and for N large enough its bounded trajectories of this finite dimensional flow equation should be very close to solutions of (2.6). Thus, up to a shift in indices, the Morse Smale complex of a_N is equal to (C^*, δ^*) so that $FH^*(E^b, E^a) \simeq H^{*-2nN}(E_N^b, E_N^a)$.

The case of a time independent Hamiltonian is essentially equivalent. The reader should note that we do not prove here that our definition of FH^* coincided with Floer's. We hope to supply such a proof in a forthcoming paper.

3 THE WEINSTEIN CONJECTURE AND RELATED RESULTS

For a long time existence of periodic solutions of an energy surface was only known for convex ([**Se**], [**We**]) then starshaped ([**R**]) hypersurfaces.

It is easy to check that these are not symplectically invariant conditions, which led A. Weinstein to seek a more general condition that would be symplectically invariant. This condition is the contact type condition which we shall state as follows

DEFINITION 3.1. *Σ is of contact type if and only if there is a vector field ξ defined in a neighborhood of Σ such that*

(i) *ξ is transverse to Σ*
(ii) *$L_\xi \omega = \omega$.*

This is obviously a symplectically invariant condition. It is also clear that if Σ is starshaped, by taking ξ to be $p\frac{\partial}{\partial p} + q\frac{\partial}{\partial q}$, then Σ is of contact type.

We should remark that if $\Sigma = H^{-1}(c)$ is a regular energy level, the trajectories of X_H on Σ do not depend on H but only on Σ: they are the integral curves of the line distribution $\ker \omega_{|\Sigma} = \mathcal{L}_\Sigma$. Such trajectories are called characteristics of Σ. Clearly periodic orbits of X_H on Σ correspond to closed characteristics. We can now state:

CONJECTURE 3.2 ([**W2**]). *If $\Sigma \subset (P, \omega)$ is a compact hypersurface of contact type, it has at least one closed characteristic.*

This conjecture was proved for $(P, \omega) = (\mathbb{R}^{2n}, \sigma)$ by the author ([**V4**]) and in several other cases in joint work with A. Floer and H. Hofer ([**F-H-V**], [**H-V1**], [**H-V2**]).

In fact as was noticed by Hofer and Zehnder ([**H-Z**]) our proof yields the more general

THEOREM 3.3. *(Quasiexistence of closed characteristics near a compact hypersurface)*

Consider an embedding of $\Sigma \times [-1, 1]$ in $(\mathbb{R}^{2n}, \sigma)$. Then for any positive ε, there is some t in $[-\varepsilon, \varepsilon]$ such that the image Σ_t of $\Sigma \times \{t\}$ has at least one closed characteristic.

Theorem 3.3 yields existence on a hypersurface of contact type as follows. Let φ_t be the flow of ξ, and set $\Sigma_t = \varphi_t(\Sigma)$. Then since $\varphi_t^* \omega = e^t \omega$, φ_t maps the characteristics of Σ on those of $\varphi_t(\Sigma)$. Hence if Σ_t has a closed characteristic, the same is true for Σ.

The proof of 3.3 goes as follows. We first construct a suitable Hamiltonian so that its periodic orbits will be closed characteristics of Σ plus some "parasites". However we can tell the former from the latter by the value of A_H: the closed characteristics correspond to negative critical values of A_H.

This last part is dealt with, using the following proposition which combines corollary 4.2 of [**V 2**] and the result of [**V 4**]

PROPOSITION 3.4. *Let $H : \mathbb{R}^{2n} \to \mathbb{R}$ be a nonnegative Hamiltonian such that*

(i) *$H(x) = (k + \frac{1}{2})|x|^2$ for $|x|$ large enough*
(ii) *$H \equiv 0$ on some open set U.*

Then A_H has at least k negative critical orbits. Moreover if the critical orbits are isolated, then for each $j \in [0, nk-1]$, there is a critical orbit x_j such that $i(x_j) \leq 2j+1 \leq i(x_j) + \nu(x_j)$.

The proposition follows from Proposition 2.2 and the fact that for some small ε and large C, we have $FH_{S^1}^{-2j+2}(E^{-\varepsilon}, E^{-C}) \neq 0$.

In fact let us write the exact homology sequences of the pairs (pt, pt), $(E^{-\varepsilon}, E^{-c})$ and the map between these sequences induced by the constant map

$$\begin{array}{ccccccc}
H^*_{S^1}(E^{-\varepsilon}, E^{-C}) & \xrightarrow{\beta_*} & H^*_{S^1}(E^{-\varepsilon}) & \xrightarrow{\gamma_*} & H^*_{S^1}(E^{-C}) & \xrightarrow{\delta^*} & \\
\uparrow & & \uparrow \lambda_* & & \uparrow \mu_* & & \\
0 & \longrightarrow & H^*_{S^1}(pt) & \longrightarrow & H^*_{S^1}(pt) & \xrightarrow{0} & \\
 & & \wr\wr & & \wr\wr & & \\
 & & \mathbb{Q}[u] & & \mathbb{Q}[u] & &
\end{array}$$

In the above diagram u is an element of degree two. Now for C large enough, it is easy to see that μ_* only depends on the behavior of H outside a compact set. Due to assumption (i) we get that $H^*_{S^1}(E^{-C})$ is isomorhic to $H^*_{S^1}(S^{2n[N-k]-1}) = H^*\mathbb{C}P^{2n[N-k]-2}$ so that $\mu_{2j} = 0$ for all $2n[N-k] \leq 2j$.

On the other hand, one can show by local considerations that for ε small enough, λ_{2j} is injective in the range $0 \leq 2j \leq 2nN$. As a result, for $2n[N-k] \leq 2j \leq 2nN$ $\gamma_{2j}(\lambda_{2j}(u^j)) = 0$ so that $0 \neq \lambda_{2j}(u^j) \in \operatorname{Im} \beta_{2j}$ and thus $H^{2j}_{S^1}(E^{-\varepsilon}, E^{-C}) \neq 0$ that we can write as $FH_{S^1}^{-2j}(E^{-\varepsilon}, E^{-C}) \neq 0$ for $0 \leq 2j \leq 2nk-2$. We can now apply proposition 2.2, and by comparing index and coindex we get a critical point with the desired value of $i(x)$. Note that in the sequel we shall call $i(x)$ the index of x. We hope that this will not cause any inconvenience It is probably Proposition 3.4 that is the most useful result. We shall actually use it again in section 4.

We now show how to infer Theorem 3.3 from 3.4(cf. [**V 2**],[**H-Z**]). Let H be defined as follows:

- $H(\Sigma_t) = \rho(t)$ with ρ as in Figure 1
- $H \equiv a$ in $\mathcal{B}(0,R) - \varphi(\Sigma \times [-1,1])$, where $\mathcal{B}(0,R)$ is a large ball containing $\varphi(\Sigma \times [-1,1])$, and a is large
- $H(x) = g(|x|)$ for $|x| \geq R$ where g is some real function such that

$$\begin{cases} g(s) = a \text{ for } s = R \\ g'(s) \leq (2k+1)s \\ g(s) \geq (k+\frac{1}{2})s^2 \text{ with equality for } s \text{ large enough.} \end{cases}$$

It is easy to check that negative critical values of A_H correspond to closed characteristics of some Σ_t (such critical value can obviously not correspond to a constant, and the conditions imposed on g prevent it to correspond to some closed characteristics on a sphere $|x| = r > R$).

Now applying 3.4 to H yields 3.3.

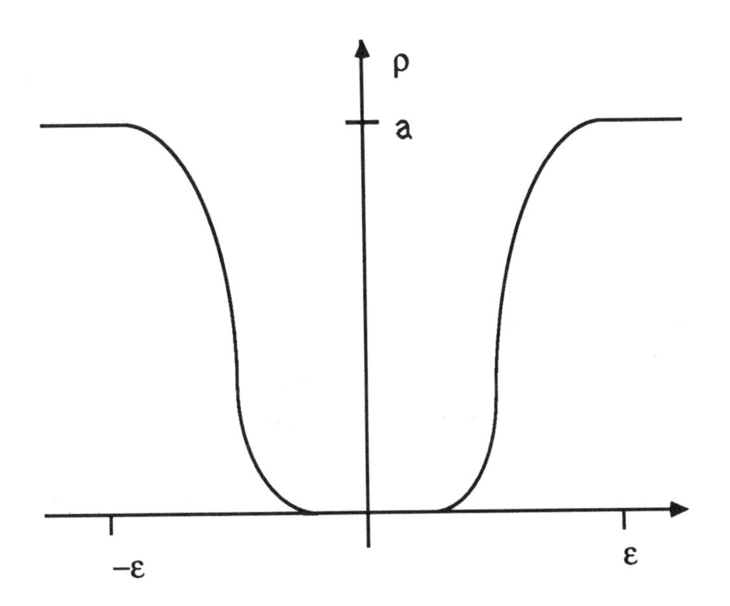

Figure 1: The function ρ

Appendix to section 3.

In fact, we prove a little more than just existence of a periodic orbit. Clearly, for every j, we get a solution of index and nullity satisfying $i \leq 2j+1 \leq i+\nu$.This can be refined, because the cohomology class that we use to get our j^{th}critical point is obtained from the preceding one by a cup-product (they satisfy a chain condition in the terminology of Lusternik-Schnirelman) and we thus get that those periodic orbits are all distinct from each other. Of course we do not claim that these orbits are *geometrically* distinct; they could just be iterates of a single orbit. However in the convex case, I.Ekeland was able to infer from this the following facts:

(1) there are always at least two geometrically distinct periodic orbits ([**E-La**])

(2) generically, there are infinitely many ([**E 1**])

This last result was extended to the starshaped case in dimensions multiples of four in [**V 3**]. We shall try to sketch the main ingredient of Ekeland's proofs; for more details we refer to [**E 1**],[**E- La**].

Let x_1 be a T periodic solution of

$$\dot{x} = X_H(x) = J\nabla H(x);$$

we would like to compute the index i_k and nullity ν_k of its kth iterate x_k as a kT periodic solution of the same equation as a function of i_1, ν_1 and the linear Poincaré map R of x_1,that is $R = d\phi_T(x_1(0)$ if ϕ_s denotes the flow of X_H. That such a formula exists is due to Bott in the framework of closed geodesics ([**Bott**]) and was extended by Ekeland for the convex case (as we show in [**V 3**]: his proof is also valid for a general Hamiltonian). The results reads as follows. Let $\{\omega_1, \ldots, \omega_m \bar{\omega}_1, \ldots, \bar{\omega}_m\}$ be the eigenvalues of R on the unit circle other than 1. We first define a function $j : S^1 \to \mathbb{Z}$ as follows

(1) $j(1) = i_1$, $j(\bar{\omega}) = j(\omega)$

(2) j is continuous except maybe at the ω_l and the $\bar{\omega}_l$

(3) If $\omega = \omega_l$ for some l, then for ω_l is of Krein type (p,q) which means that the signature of the form $i(Jz,z)$ on the complex space $\ker(R_{\mathbf{C}} - \omega\, Id_{\mathbf{C}^{2n}})$ is (p,q) (where $R_{\mathbf{C}}$ is the complexification of R) :

$$j(\omega^+) - j(\omega^-) = p - q$$

here $j(\omega^{\pm}) = \lim_{\varepsilon \to 0^{\pm}} j(e^{i\varepsilon}\omega)$

(4) $j(1^{\pm}) \geq i_1 + n + 1$ and the equality holds if x_1 is nondegenerate.

Then the following formula holds

$$i_k = \sum_{\omega^k = 1} j(\omega)$$

In the convex case one can show, using this formula, that $i_{k+1} - i_k \geq 2$ and that equality cannot hold for all k's. As a result, using proposition 3.4 it is easy to see that all the solutions cannot be iterates of a single one, i.e. there must be at least two geometrically distinct periodic orbits. To prove that there are infinitely many orbits, one shows that if there were finitely many, then generically the $i_{k+1} - i_k$ have a "random behaviour", so that in the set of i_k for all k and all solutions, there will be holes of length at least 3 in this set which contradicts our result.

An other important question, the minimality of the period for superquadratic convex Hamiltonians was also dealt with by Ekeland and Hofer in [**E-H 1**], by using index theory. We refer to it and to the forthcoming book by Ekeland for related questions.

4 APPLICATIONS: CAPACITIES AND LAGRANGIAN EMBEDDINGS

Consider again H as in Proposition 3.4. Then the map $E^{-\lambda} \to pt$ induces a map $\mathbb{Q}[u] = FH^*_{S^1}(pt) \to FH^*_{S^1}(E^{-\lambda})$ which is injective for λ small enough, and vanishes in dimensions $0 \leq 2j \leq 2k - 2$ if λ is large. We can then define $c_j(H)$ to be the smallest C for which the cohomology class $\lambda_{2n[N-k]+2j}(u^j)$ defined in the previous section is in the image of β_*.

DEFINITION 4.1. *$c_j(H)$ is the jth Ekeland-Hofer capacity of H.*

Since $H_1 \leq H_2$ implies $A_{H_1} \leq A_{H_2}$ hence $E_2^{-C} \subset E_1^{-C}$ (E_i^{-C} corresponds to H_i) and thus $c_j(H_2) \leq c_j(H_1)$.

We can now define capacities for sets by

DEFINITION 4.2. *For U a compact set, define*

$$c_j(U) = \inf\{c_j(H) \mid H \in \mathcal{F}_k(U) \text{for some } k \in \mathbb{Z}\}$$

where $\mathcal{F}_k(U) = \{H \in C^\infty(\mathbb{R}^{2n}) \mid H \equiv 0$ *on* U *and* $H(x) = \left(k + \frac{1}{2}\right)|x|^2$ *for* $|x|$ *large enough*$\}$. *For U noncompact, define*

$$c_j(U) = \sup\{c_j(V) \mid V \text{ compact}, V \subset U\}.$$

It is not our purpose here to study the properties of capacities; we refer the reader to [**E-H 2**]. The main use of capacities follows from the obvious fact that

PROPOSITION 4.3. *If $U \subset V$ $c_j(U) \leq c_j(V)$.*

Thus capacities are a useful and simple tool to deal with embedding problems, i.e. to decide when can some symplectic open set U be embedded in V.

It is also easy to see that, at least in good cases, the capacity will be represented as the area of some closed characteristics of the boundary of U. We actually have the following theorem of [**E-H 2**]

THEOREM 4.4. *(representation theorem)*([**E-H 2**], [**H-Z**]) *Let Σ be the boundary of U; then if Σ is of restricted contact type then $c_j(U)$ is the area of some closed characteristics on Σ.*

Restricted contact type means that there is a one form α on $\mathbb{R}^{2n}$ such that $d\alpha = \omega$ and α induces a contact form on Σ

Note that if Σ is starshaped, it is of restricted contact type, as we see by taking $\alpha = \frac{1}{2}(pdq - qdp)$. The proof can be sketched as follows. Take a cofinal sequence of Hamiltonians in $\mathcal{F}(U) = \bigcup \mathcal{F}_k(U)$. Then for each H_m in the sequence, $c_j(H_m)$ is represented by some closed characteristics γ_m of Σ of index $2j+1$. The hypotheses on Σ imply that the length of a characteristic loop, γ_m, is bounded by $\int_{S^1} \gamma_m^* \lambda$. On the other hand, $c_j(H_m) = \int_{S^1} \gamma_m^* [\lambda - H_m(\gamma_m)]$ One can show that $H_m(\gamma_m)$ is small so that the length ofγ_m gets bounded by $c_j(H_m) + \epsilon$ and we then apply the Ascoli-Arzela compactness to conclude that the γ_m converge to γ necessarily a characteristic of Σ with $\int_{S^1} \gamma^* \lambda = c_j(\Sigma)$

REMARKS. *It is also proved in [**E-H 2**] that* $c_1(U) = c_1(\Sigma)$

As a result, one can prove the following theorem of Gromov:

THEOREM 4.5. *There can be no embedding of* $B^{2n}(R)$ *into* $B^2(R') \times \mathbb{R}^{2n-2}$, *unless we have* $R \leq R'$

The above theorem has a lot of nice consequences as Ekeland, Hofer and indipendently Eliashberg had noticed, the most striking being

THEOREM. *(C^0 closure of symplectic diffeomorphisms) Let* ψ_n *be a sequence of symplectic diffeomorphism converging for the* C^0 *topology to some map* ψ. *Then if* ψ *is* C^1,*it is in fact a symplectic diffeomorphism.*

In [**E-H 2**] it is proved that any C^1, capacity preserving map (i.e. such that $c_1(\psi(U)) = c_1(U)$) must be symplectic. Since capacities are C^0 continuous, that is are continuous for the Hausdorff distance, this implies the above result. As a consequence of the theorem, one can speak of symplectic topology at the C^0 level, by defining the group $\mathcal{S}_\omega$ of symplectic homeomorphism as the set of those homeomorphism which preserve the first capacity and go on defining symplectic topological manifolds as a topological manifold with atlas in $\mathcal{S}_\omega$. It is not really clear what these are nor is it clear whether requiring to preserve higher capacities i.e. c_j for $j > 1$ really gives a different group of symplectic homeomorphisms. We shall not say more about capacities since several survey papers on the subject are about to appear.

We now turn to the problem of lagrangian embedding of the torus. Let L be an n-dimensional manifold, and j an embedding of L in $\mathbb{R}^{2n}$. We call j lagrangian if the pull-back of σ by j vanishes (i.e. $j^*\sigma = 0$). Since $\sigma = d\lambda$ where λ is the Liouville form $\sum_{i=1}^n x_i \wedge dy^i$ the pull-back of λ is a closed form on L. If this form is exact, we shall say that the embedding is exact. A celebrated theorem by M. Gromov ([**G1**]) can be stated as follows:

THEOREM 4.7. *There is no exact lagrangian embedding in* $(\mathbb{R}^{2n}, \sigma)$

Gromov's proof mainly uses properties of Cauchy-Riemann elliptic equations. We here would like to give a proof of a somehow more precise result, at least when L is the n-torus. Our result states:

PROPOSITION 4.8. *Let L be the n-torus, and j as above. Then there exists a closed loop γ on the torus such that*

(1) $(j^*(\lambda), \gamma) > 0$
(2) $2 \leq (\mu(j), \gamma) \leq 2[\frac{n+1}{2}]$

(This is due to Bennequin ([**Be**]) for $n = 2$.) In (2), $\mu(j)$ designates the Maslov class of the embedding, $(\mu(j), \gamma)$ measures the rotation of the tangent plane to the torus along the curve γ. We refer to [**Ar**] for the exact definition.This is the new fact with respect to theorem 4.6. Let us remark that the Maslov class is invariant by regular homotopy of lagrangian immersions. On the other hand, usual immersions of the torus are classified, up to regular homotopy, by the their normal bundle (in dimension greater than 2). Since for a lagrangian immersion the tangent and normal bundles are isomorphic (this follows from a result by A. Weinstein stating that $j(L)$ has a tubular neighbourhood symplectomorphic to a neigbourhood of the zero section in $(T^*L, dp \wedge dq)$),there is no a priori obstruction to embedding the n-torus with any even Maslov class. On the other hand, our proposition implies that there is such an obstruction: for instance a two-torus cannot be embedded with Maslov class $(0,4) \in H^1(T^2, \mathbb{Z}) = \mathbb{Z}^2$ since no loop will have Maslov number equal two. We now give the ideas of the proof of our proposition. The main idea is that, using the above mentioned theorem by A. Weinstein, one gets an embedding of a small sphere bundle $\Sigma(\epsilon) = \{(q,p) \in T^*T^n \mid |p| = \epsilon\}$, in $\mathbb{R}^{2n}$ and to remark that $\Sigma(\epsilon)$ is a contact manifold whose characteristic flow corresponds to the geodesic flow of T^n for the standard flat metric. As a result, the closed characteristics of the flow project onto the closed geodesics of the torus, which are well known. We can now use our machinery of section 3, and get a closed characteristic of positive action and index one on $\Sigma(\epsilon)$. The index of a closed characteristic of $\Sigma(\epsilon)$ can be computed as the sum of the index of the corresponding geodesic and the Maslov number of this geodesic minus n; we write:

$$i(x) = i_g(q) + (\mu(j), q) - n \text{ for } x = (q,p)$$

where $i_g(q)$ is the Morse index of q as a geodesic.

The above equality is quite natural: both indices $i(x)$ and $i_g(q)$ can be defined as rotation numbers of lagrangian planes, the difference between them, is that $i(x)$ measures the rotation with respect to a fixed plane in $\mathbb{R}^{2n}$ while $i_g(q)$ measures the rotation of the same lagrangian plane , but this time with respect to the vertical plane in T^*T^n. It is now clear that they will differ exactly by the rotation of the vertical plane of T^*T^n with respect to a fixed plane in $\mathbb{R}^{2n}$, that is by definition the Maslov number of the curve. Note now that all geodesics on T^n have nullity $n-1$ since they belong to an n dimensional nondegenerate manifold. Thus according to proposition 3.4, there is at least one periodic solution x such that $i(x) \leq 1 \leq i(x)+n-1$ or else

$i_g(q)+(\mu(j),q)-n \leq 1 \leq i_g(q)+(\mu(j),q)-1$ that is, since the geodesics of T^n all have index zero,

$$2 \leq (\mu(j),q) \leq n+1$$

We still have to show that $(j^*(\lambda),\gamma) > 0$; for this we use the fact that the action functional is positive on x. We can break down the functional in two parts:

$$A_H(x) = \int_0^T [x^*(\lambda) + H(x)]\, dt = \int_0^T [p\dot{q} + H(x)]\, dt + \int_0^T q^*(j^*(\lambda))$$

since the form $\alpha = pdq - j^*(\lambda)$ defined in a neighbourhood of T^n is closed and thus satisfies $x^*\alpha = q^*\alpha = q^*(j^*(\lambda)$. One can show that the first term goes to zero with ϵ provided the Hamiltonian is chosen wisely. Thus we must have $\int_0^T q^*(j^*(\lambda)) = (j^*(\lambda),q) \geq 0$. We now just have to prove strict inequality. If one had $(j^*(\lambda),q) = 0$ for all q with Maslov class in the proper interval, we could then slightly move j so that $(j^*(\lambda),q)$ actually becomes negative on these loops without changing their Maslov number, thus yielding a contradiction. This concludes the section.

5 AN EXISTENCE MECHANISM

In this section, we shall work on a general *compact* symplectic manifold (P, ω). Our aim is to extend the results of section 3 to certain classes of manifolds.

Note that here quasiexistence often makes more sense than contact type condition, e.g. $P \times S^1 \subset (P \times S^2, \omega \oplus \sigma)$ is not of contact type (since ω is not exact). However, a neighborhood of $P \times S^1$ is foliated by hypersurfaces with the same characteristic flow (take the images of $P \times S^1$ under the radial flow) so that quasiexistence implies existence (see in the appendix a suggested definition of generalized contact type).

In a joint paper with A. Floer and H. Hofer ([**F-H-V**]) the following result was proved.

PROPOSITION 5.1. *Let (P, ω) be a compact symplectic manifold such that $[\omega]\pi_2(P) = 0$. Then for any compact hypersurface in $P \times \mathbb{C}$, quasiexistence of periodic orbits holds.*

To prove this result we reduce the problem to finding period solutions of a Hamiltonian system with negative action as we did in section 3. Then existence of such orbits is proved using the holomorphic curve technique of Gromov and Floer ([**Gr**], [**Fl1**], [**Fl2**]).

We now would like to explain a technique developed in [**H-V2**] which also relies on holomorphic curves.

To begin with, we define an almost complex structure on (P, ω) to be an automorphism J of TP such that

(i) $\omega(J\xi, J\eta) = \omega(\xi, \eta)$

(ii) $g(\xi, \eta) = \omega(J\xi, \eta)$ is a Riemannian metric on P.

Almost complex structures always exist on symplectic manifolds, and are unique up to homotopy.

In the sequel we shall assume (P, ω) to be endowed with such a structure.

It is then clear what holomorphic maps are: namely $f : (S, J_0) \to (P, J)$ is holomorphic if and only if $Df \circ J_0 = J \circ Df$. Note that if f is a holomorphic map of a Riemann surface, then $f^*\omega$ is positive on S. As a result, the condition $[\omega]\pi_2(P) = 0$ implies there are no holomorphic spheres in P, since for such a sphere $\int_{S^1} f^*\omega > 0$ and $< [\omega], f_*[S^2] > \neq 0$.

As opposed to **[F-H-V]**, the technique in **[H-V2]** *requires* the existence of holomorphic spheres, but in a somewhat controllable way. We shall explain our method with the assumption that for some pair of points x_0, y_0 there is a unique (up to conformal transformation) holomorphic sphere S_0 in P through x_0 and y_0 which is nondegenerate and has least area among all nontrivial holomorphic spheres.

PROPOSITION 5.2. *Under the above assumptions, any hypersurface Σ of P such that x_0 is in the inside of Σ, y_0 is in the outside of Σ, satisfies quasiexistence.*

REMARK 1: This applies for instance to $P = \mathbb{C}P^n$.

REMARK 2: If on P there exist a pair (x_0, y_0) satisfying the hypothesis of Proposition 5.1 for a given almost complex structure, then the hypothesis will be satisfied for every pair (x, y) in P, possibly with a different almost complex structure. This follows easily from the transitivity of the group of symplectic diffeomorphisms on pairs of points. As a result we can weaken the hypothesis of 5.2 by just assuming

- existence of a pair (x_0, y_0) with a unique minimal holomorphic sphere
- that $P - \Sigma$ has two connected components.

We now explain the main ideas of the proof. First of all, we reduce the problem to existence of nonconstant periodic orbits for a Hamiltonian H which is constant outside a neighborhood of Σ. What we actually prove is worth stating in its own right.

PROPOSITION 5.3. *Let H be a Hamiltonian on P such that $H \equiv 0$ in a neighborhood of x_0, $H \equiv 1$ in a neighborhood of y_0, and such that H has no critical values in $]0,1[$. Then $\dot{x} = X_H(x)$ has a nonconstant periodic solution of period $T < \langle \omega, S_0 \rangle$.*

To begin with, let us recall the interpretation of holomorphic spheres as flow trajectories of the action functional.

Although A_H is only well defined if $[\omega]\pi_2(P) = 0$, DA_H is always defined as

$$DA_H(x)\xi = \int_{S^1} [g(J\dot{x}, \xi) + dH(x)\xi] d\theta$$

and the L^2 gradient flow of $-A_H$ is given by

$$\frac{\partial}{\partial t}x_t = -J\frac{\partial}{\partial \theta}x_t - \nabla H(x_t). \tag{5.4}$$

We set $u(t,\theta) = x_t(\theta)$ so that (5.3) becomes

$$\begin{aligned} \overline{\partial} u &= \frac{\partial}{\partial t}u + J\frac{\partial}{\partial \theta}u = -\nabla H(u), \text{ or} \\ \overline{\partial} u &= -\nabla H(u). \end{aligned} \tag{5.5}$$

Equation (5.5) is the inhomogeneous Cauchy-Riemann equation. Now, if we look at a flow trajectory connecting two critical points of A_H, say x_- and x_+, we must add to (5.5) the boundary conditions

$$\lim_{t\to+\infty} u(t,\theta) = x_+(\theta), \ \lim_{t\to-\infty} u(t,\theta) = x_-(\theta). \tag{5.6}$$

In particular, if $H \equiv 0$ and x_+, x_- are constants, we can extend u—originally defined on the cylinder $\mathbb{R} \times S^1$—to the sphere S^2, by conformally identifying $\mathbb{R} \times S^1$ to $S^2 - \{0, \infty\}$.

Thus our hypothesis just says that there is a unique "flow trajectory of $-A_0$" (remember that only DA_H is well defined) connecting the constant map at x_0 with the constant map at y_0.

Note that if we fix the homotopy class of a map u satisfying (5.6) we can define $A_H(x_+) - A_H(x_-)$ as

$$\int_{\mathbb{R}\times S^1} u^*\omega - \int_{S^1} [H(x_+(\theta)) - H(x_-(\theta))]d\theta.$$

Of course if there is a flow trajectory of $-A_H$ connecting x_+ and x_-, we must have $A_H(x_+) - A_H(x_-) > 0$.

For $H \equiv 0$ $x_+ \equiv x_0, x_- \equiv y_0$, we find $\int_{\mathbb{R}\times S^1} u^*\omega > 0$; this just means that a holomorphic sphere has positive area.

We now replace A_0 by A_H and look again at connecting trajectories between x_0 and y_0 in the same homotopy class as u_0. A necessary condition for such a trajectory to exist is that $A_H(x_0) - A_H(y_0) > 0$ when the difference is computed along u, that is

$$\int_{\mathbb{R}\times S^1} u^*\omega - \int_{S^1} [H(y_0) - H(x_0)]d\theta > 0$$

or

$$\langle \omega, S_0 \rangle - T[H(y_0) - H(x_0)] > 0 :$$

since by assumption $H(y_0) - H(x_0) = 1$, this last inequality can be written $T < \langle \omega, S_0 \rangle$.

As a result for $T = \langle \omega, S^2 \rangle$ there cannot be any such trajectory.

We now consider for λ in $[0,1]$ the set of trajectories of the flow for $A_{\lambda H}$ in the homotopy class of u_0.

Using a degree argument, the change in the algebraic number of trajectories betwen $\lambda = 0$ and $\lambda = 1$ can only be accounted for by a lack of compactness of the set of solutions as $\lambda_n \to \lambda$ in $[0,1]$.

Finally such a lack of compactness could only be caused by

- bubbling off of a holomorphic sphere. We use energy inequality to show that the assumption that S_0 has minimal area prevents this phenomena.
- splitting of the trajectory into other connecting trajectory involving at least a third critical point of A_H. One can show that this critical point of $A_{\lambda H}$ necessarily corresponds to a nonconstant T-periodic solution of $\dot{x} = \lambda X_H(x)$, hence by rescaling to a nonconstant λT-periodic solution of $\dot{x} = X_H(x)$ as promised.

We would like to point out that our technique can be extended in order to replace x_0 and y_0 by submanifolds of P. As a result, we can prove quasiexistence for any $\Sigma \subset P \times S^2$ such that $\Sigma \cap P \times \{\infty\} = \emptyset$; that is $\Sigma \subset P \times \mathbb{C}$, and thus get a new proof of Proposition 5.1.

Also, working in $P \times S^2$ allows us to prove

PROPOSITION 5.7. *Let ψ_t be the flow of a time dependent Hamiltonian $H(t,p)$ on (P,ω). Then ψ_1 has at least one periodic point, i.e. there is a point p_0 in P such that $\psi_1^n(p_0) = p_0$. Moreover n can be assumed to satisfy $0 < n \leq \nu\ (\|H\|_{C^0}, P)$.*

REMARK: $\nu(\|H\|_{C^0}, P)$ actually depends only on $\|H\|_{C^0}$ and the smallest area of a holomorphic sphere in P for some almost complex structure. In particular, if we call $a(P,J)$ this area, we can assume

$$\nu(\|H\|_{C^0}, P) \leq \max\{\|H\|_{C^0} / \sup_J a(P,J), 1\}.$$

Note that $\sup_J a(P,J)$ is equal to Gromov's width (cf. [**G2**]).

Appendix to section 5.

We would like to suggest the following definition:

DEFINITION 5.8. *Let Σ be an hypersurface in (P, ω); we shall say that Σ is of generalized contact type if there is a vector field ξ defined in a neighbourhood of Σ such that:*

(1) *ξ is transverse to Σ*
(2) *$ker\, \omega_{|\Sigma} \subset ker\, L_{\xi}\omega_{|\Sigma}$.* □

Condition (2) implies that if ϕ_t is the flow of ξ, then $ker\, \omega_{|\phi_t(\Sigma)} = \phi_t(ker\, \omega_{|\Sigma})$. Clearly (2) is weaker than the usual condition $L_{\xi}\omega = \omega$. With the above definition, $P \times S^1 \subset P \times S^2$ is of generalized contact type, as easily checked by using the vector field $\xi = (0, r\frac{\partial}{\partial r})$. Also all our proofs of the Weinstein conjecture work if we replace therein contact type by generalized contact type.

6. Final comments and conjectures

In this paper we tried to convey the feeling that periodic orbits of autonomous Hamiltonian systems are important in several ways, and to show some new techniques that have been successful in dealing with them. We saw in section 4 that, as far as symplectic geometry is concerned, periodic orbits obviously appear in the Weinstein conjecture, hence in the structure of contact manifolds, but also, less obviously when dealing with lagrangian embeddings. In fact these can be considered as the two extremes of the following conjecture that we would like to give as a conclusion:

CONJECTURE. *Let K^{n+p} be a "regular" compact embedded coisotropic submanifold of $\mathbb{R}^{2n}$, and let $\mathcal{K}^{n-p}$ be its coisotropic foliation: then if λ is the Liouville form on $\mathbb{R}^{2n}$ there exists a leaf $\mathcal{K}_x^{n-p}$ such that λ induces on it a non-zero cohomology class. As a result, there is at least one non-simply connected leaf.*

We see that if K is of codimension one, then $\mathcal{K}^{n-p}$ is just the characteristic foliation; if we interpret the word "regular" as meaning of generalized contact type, then the conjecture essentially boils down to the Weinstein conjecture(the only non-simply connected one dimensional manifold is the circle!). On the other hand if K is of dimension n, then it is a Lagrangian submanifold, and if we interpret the condition "regular" to be empty, then the conjecture boils down to the non-existence of exact Lagrangian embeddings.

Let us point out that if the reader does not know the meaning of "regular" the author doesn't either, except in the two cases mentioned.

On the other hand, this conjecture should be a motivation for a unified proof of these two results, in the hope that such a proof would lead to a proof of the above conjecture.

References

[**A-Z 1**] Amann, H., Zehnder, E., *Nontrivial solutions for a class of nonresonance problems and applications to nonlinear differential equations*, Ann. Scuola Norm. Pisa Cl. Sci.(4) **7** (1980), 539-603.

[**A-Z 2**] Amann, H., Zehnder, E., *Periodic solutions of asymptotically linear Hamiltonian systems*, Manuscripta Math. **32** (1980), 149-189.

[**Ar**] Arnold, V.I.,, *Appendix*, in "Maslov, V.P., Théorie des perturbations et méthodes asymptotiques (traduit du russe)," Dunod- Gauthier-Villars, Paris,, 1972.

[**Be**] Bennequin,D., "Communication au congrès de Géométrie Symplectique de La Grande Motte," 1988.

[**B-L-M-R**] Berestycki, H.; Lasry,J.M.; Mancini,G. and Ruf, B., *Existence of Multiple Periodic Orbits on Starshaped Hamiltonians Systems*, Comm. Pure Appl. Math. **38** (1985), 253-289.

[**Bor**] Borel, A. "Seminar on transformation groups," Annals of Math. Studies n°46, Princeton University Press, New-York, 1960.

[**Bott**] Bott,R., *On the iteration of closed geodesics and the Sturm intersection theory*, Commun. in Pure and Appl. Math. **9** (1956), 176-206.

[**C**] Chaperon, M., *Une idée du type géodésiques brisées pour les systèmes hamiltoniens*, Comptes Rendus Acad. Sci. Paris **298** (1984), 293-296.

[**Ch**] Chazarain, J., *Formule de Poisson pour les variétés Riemanniennes*, Invent. Math. **24** (1974), 65-82.

[**Cl**] Clarke,F., *Periodic solutions to Hamiltonians inclusions*, J. Differential Eqations **40** (1980), 1-6.

[**Cl-E**] Clarke,F., Ekeland,I., *Hamiltonian Trajectories having Prescribed Minimal Period*, Comm. Pure Appl. Math. **37** (1980), 103-116.

[**Co**] Conley,C.C., "Isolated Invariant Sets and their Morse Index," C.B.M.S. Reg. Conf. Series in Math. n° 38, Amer. Math. Soc., Providence,R.I., 1978.

[**D-G**] Duistermaat,J., Guillemin,V.,W., *The spectrum of Positive Elliptic Operators and Periodic Bicharacteristics*, Invent. Math. **29** (1975), 39-79.

[**E**] Ekeland,I., *Une théorie de Morse pour les systèmes Hamiltoniens convexes*, Ann. Inst. H. Poincaré Anal. Non Linéaire **1** (1984), 19-78.

[**E-La**] Ekeland,I.,Lassoued,L., *Multiplicité des trajectoires fermées de systèmes hamiltoniens convexes*, Ann. Inst. H. Poincaré Anal. Non Linéaire (1987), 307-336.

[**E-H 1**] Ekeland,I.,Hofer, H., *Periodic Solutions with prescribed minimal period for convex autonomous systems*, Invent. Math. (1985), 155-188.

[**E-H 2**] Ekeland,I.,Hofer, H., *Hamiltonian dynamics and symplectic geometry*, preprint.

[**Fl 1**] Floer,A., *The unregularized Gradient Flow of the symplectic Action*, Comm. Pure Appl. Math. (to appear).

[**Fl 2**] Floer,A., *Morse Theory for Lagrangian Intersections*, J. Differential Geometry (to appear).

[**F–H–V**] Floer, A. Hofer,H. Viterbo,C., *The Proof of Weinstein Conjecture in $P \times \mathbb{C}$.*

[**G 1**] Gromov,M., *Pseudo Holomorphic Curves on almost complex Manifolds*, Invent. Math. **82** (1985), 307–347.

[**G 2**] Gromov,M., *Soft and Hard Symplectic Geometry*, in "Proceedings of the International Congress of Mathematicians 1986," 1987, pp. 81–98.

[**H-V 1**] Hofer,H., Viterbo,C., *The Weinstein conjecture for Cotangent Bundles and related Results*, Ann. Scuola Norm. Pisa Cl. Sci. (4) (to appear).

[**H-V 2**] Hofer,H.,Viterbo,C., *The Weinstein conjecture for Compact manifolds and related Results*, to appear.

[**H-Z**] Hofer,H., Zehnder,E., *Periodic Solutions on Hypersurfaces and a Result by C. Viterbo*, Invent. Math. **90** (1987), 1-9.

[**M**] Milnor, J., "Lectures on the h-cobordism theorem," Princeton University Press, Princeton, N.J., 1965.

[**R**] Rabinowitz,P.H., *Periodic solutions of Hamiltonians systems*, Comm. Pure Appl. Math. **31** (1978), 157-184.

[**Se**] Seifert, H., *Periodische Bewegungen mechanischer Systeme*, Math. Zeitschr. **51** (1948), 197-216.

[**V 1**] Viterbo,C., *Intersections de sous- variétés lagrangiennes, fonctionelles d'action et indice des systèmes hamiltoniens*, Bull. Soc. Math. France **115** (1987), 361–390.

[**V 2**] Viterbo,C., *A proof of Weinstein's conjecture in $\mathbb{R}^{2n}$*, Ann. Inst. H. Poincaré Anal. Non Linéaire **4** (1987), 337-356.

[**V 3**] Viterbo,C., *Equivariant Morse theory for starshaped Hamiltonian systems*, Trans. Amer. Math. Soc. (to appear).

[**V4**] Viterbo,C., *Indice de Morse des points critiques obtenus par minimax*, Ann. Inst. H. Poincaré Anal. Non Linéaire **5** (1988), 221-225.

[**V5**] Viterbo,C., *A New Obstruction to embedding Lagrangian tori*, (in preparation).

[**We 1**] Weinstein, A., *Periodic Orbits for Convex Hamiltonian Systems*, Ann. of Math. **108** (1978), 507-518.

[**We 2**] Weinstein, A., *On the Hypotheses of Rabinowitz' periodic orbit theorem*, J. Differential Equations **33** (1979), 353-358.

[**We 3**] Weinstein, A., "Lectures on symplectic manifolds," C.B.M.S. Reg. Conf. Series in Math., n° 29, Amer. Math. Soc., Providence, R.I., 1979.

Critical Points in the Presence of Order Structures

Krzysztof Wysocki *
Department of Mathematics
Rutgers University
New Brunswick, NJ 08903

1 Introduction

In an interesting paper [2], Hofer proved the existence of multiple critical points for functionals whose potential operators preserve an order structure. His results also show that the existence of this additional structure provide more information about the set of critical points. In this paper we give an extension of some of these ideas. We show that under certain circumstances the existence of local minima of Φ ordered in a special way "forces" a functional Φ to have many additional critical points. Finally, we show how these abstract results apply to a concrete situation. For detailed proofs we refer the reader to [4].

We denote by H a real Hilbert space and by $\Phi : U \subseteq H \to R$ a C^1 -functional on an open subset U of H.
We put

$$\begin{aligned} Cr(\Phi, C, d) &:= \{u \in C \;:\; \Phi(u) = d,\; \Phi'(u) = 0\} \\ Cr(\Phi, C) &:= \bigcup_{d \in R} Cr(\Phi, C, d) \end{aligned}$$

and

$$\Phi^d := \Phi^{-1}((-\infty, d]) \;,\; \Phi_d := \Phi^{-1}([d, \infty)).$$

If C is a subset of U, we say that Φ satisfies Palais - Smale condition on C, $(PS)_C$, if for every sequence $\{u_n\} \subseteq C$ such that $\{\Phi(u_n)\}$ is bounded and

*Current affiliation: Center for Dynamical Systems and Nonlinear Studies, School of Mathematics, Georgia Institute of Technology, Atlanta, Georgia

$\Phi'(u_n) \to 0$ there is a convergent subsequence $u_{n_k} \to u \in C$.
For an isolated critical point $u \in C$ of $\Phi \in C^1(U, R)$ we can define the Poincaré series of u (relative to C) by

$$P_{\Phi,C,u}(t) = \sum_{i=0}^{\infty} \dim H^i(\Phi^d \cap C \cap W_u; (\Phi^d \cap C \setminus \{u\}) \cap W_u)$$

where $d := \Phi(u)$ and W_u is an open neighborhood of u so that $Cr(\Phi, C) \cap W_u = \{u\}$.
Here $H^i(A, B)$ stands for ith singular cohomology group of the couple of topological spaces (A, B).
For example, if $u \in C$ is a local minimum of Φ then $P_{\Phi,C,u}(t) = 1$.
We note that if u is in the interior of C then the above definition agrees with the standard one; in this case we write $P_{\Phi,u}(t)$. We will call a critical point $u \in C$ trivial if its Poincare series $P_{\Phi,C,u}(t) = 0$; otherwise u is nontrivial.

We employ the standard notations of ordered spaces; if P is a positive cone in H we write $x \ll y$ iff $y - x \in P$; $x < y$ iff $x \leq y$ but $x \neq y$. If $int\ P \neq \emptyset$ $x \ll y$ means $y - x \in int\ P$. An order interval $[x, y]$ is the set of points z such that $x \leq z \leq y$, and x, y are called comparable if $x - y \in P \cup -P$.
Finally, an operator $L : H \to H$ is order preserving if $x \leq y$ implies $Lx \leq Ly$ and strongly order preserving if $x < y$ implies $Lx \ll Ly$.

We impose following condition :

(Φ) (H, P) is an ordered Hilbert space, the cone P has nonempty interior, $U \subseteq H$ is open and convex. $\Phi \in C^2(U, R)$ with a gradient $\Phi' = I - K$ with K being compact and strongly order - preserving. If $u \in Cr(\Phi, U)$ then $K'(u)$ is strongly order - preserving and that for any $u, v \in Cr(\Phi, U)$ Φ satisfies $(PS)_{[u,v]}$.

Let $\Sigma := \{u \in U\ ; u$ is local minimum of $\Phi\}$. With Σ we associate an abstract graph (Σ, Γ) where Γ is the set of edges defined by

$$\Gamma = \{(u, v) \in \Sigma \times \Sigma\ ; u \ll v \text{ and there is no } w \in \Sigma \text{ such that } u \ll w \ll v\}.$$

For a given u let $\mathcal{V}$ be the set of vertices of a standard n - cube I^n i.e $\mathcal{V} = \{(\alpha_1, \ldots, \alpha_n)\ ; \alpha_i = 0 \text{ or } 1\}$. We introduce ordering on $\mathcal{V}$ by

$$(\alpha_1, \ldots \alpha_n) \leq (\beta_1, \ldots, \beta_n) \Leftrightarrow \alpha_i \leq \beta_i, i = 1, \ldots n.$$

Definition 1 *Let $\Sigma^n \subseteq \Sigma$ be such that if $u, v \in \Sigma^n$ and $w \in \Sigma$ and $u \ll w \ll v$ or $v \ll w \ll u$ then $w \in \Sigma^n$.*
We say that Σ^n is an n - cube if there is an isomorphism $f: \Sigma^n \to \mathcal{V}$ which satisfies

$$v \ll u \Leftrightarrow f(v) < f(u).$$

If $k \leq n, \Sigma^k \subseteq \Sigma^n$ is a k - subcube of Σ^n if $\Sigma^k = f^{-1}(\mathcal{V}^k)$ where $\mathcal{V}^k$ is the set of vertices of some k dimensional face of a standard n - cube I^n.

With an n - cube Σ^n we associate an order interval $[p, q]$where p and q are the smallest and the greatest elements of Σ^n.
After these preliminaries we can state the main results :

Theorem 1 *Suppose that condition (Φ) is satisfied and that Σ^n is an n -cube. Let $C = [p, q]$ be an order - interval corresponding to Σ^n and $\{C_i\,; 1 \leq i \leq 2n\}$ order - intervals corresponding to all $(n-1)$ - subcubes of Σ^n. Assume that Φ is bounded from below on C, that $Cr(\Phi, C)$ is finite and if $u \in Cr(\Phi, C)$ then $\dim \ker \Phi''(u) \leq 1$..*
Then there is an odd number of critical points in C and that number is at least equal to 3^n. Moreover, if $C^0 = C \setminus \bigcup_{i=1}^{2n} C_i$ then C^0 contains an odd number of nontrivial critical points which are not local minima of Φ.

If we denote by η_k the cardinality of the set $\{\Sigma_0\,; \Sigma_0$ is a k - cube in $\Sigma\}$ then we have :

Theorem 2 *Suppose that (Φ) is satisfied, $Cr(\Phi, U)$ is finite and that Φ is bounded from below on any $[u, v]$; $u,v \in \Sigma$.*
If $u \in Cr(\Phi, U)$ we assume that $\dim \ker \Phi''(u) \leq 1$. Let $l := |\Sigma| =$ the cardinality of Σ and k the largest integer such that $2^k \leq l$.
Then Φ has at least $\sum_{i=1}^{k} \eta_i$ nontrivial critical points.

As an application of the last result we consider

$$\text{(1)} \qquad \begin{aligned} \epsilon u'' + u(1-u)(u - a(t)) &= 0 \\ u'(0) = u'(1) &= 0 \end{aligned}$$

where $\epsilon > 0$ and a is a C^1 - function $[0,1] \to (0,1)$ such that $a(t) \neq 1/2$ if $t = 0, 1$ and $a'(t) \neq 0$ whenever $a(t) = 1/2$.

Theorem 3 *If k is the number of solutions of $a(t) = 1/2$ and ϵ is small then there are at least $\frac{2^{k+3}+(-1)^k}{3}$ solutions of Equation (1).*

2 Proofs of main results.

The main tool in proving Theorem 1 is following :

Proposition 1 *(Morse Equation) Let $\Phi \in C^2(u, R)$, $C \subseteq U$ is convex and closed. Assume that Φ satisfies $(PS)_C$ and its gradient Φ' has a form $I - K$ with $K(C) \subseteq C$. Let $a < b$ be regular values of Φ on C and assume that $Cr(\Phi, C, (a, b))$ is finite, and if $u \in Cr(\Phi, C, (a, b))$ then $\dim H^i(\Phi^d \cap C, \Phi^d \cap C \setminus \{u\}) < \infty$, for all $i \in Z^+$.*
Then there is $Q(t) \in Z^+[[t]]$ such that

$$\sum_{u \in Cr(\Phi, C, (a,b))} P_{\Phi,C,u}(t) = P(\Phi^b \cap C, \Phi^a \cap C) + (1 + t)Q(t)$$

where $P(\Phi^b \cap C, \Phi^a \cap C) = \sum_{i \geq 1} \dim H^i(\Phi^b \cap C, \Phi^a \cap C)$.
If in addition Φ is bounded from below on C and $Cr(\Phi, C)$ is finite then

$$\sum_{u \in Cr(\Phi, C)} P_{\Phi,C,u}(t) = 1 + (1 + t)Q(t).$$

We would like to know how to calculate Poincaré series of some critical points. It is a well known result that if $\Phi \in C^2(U, R)$, $\Phi' = I - K$ with K compact and u is a critical point of Φ such that $\ker \Phi''(u) = 0$, then $P_{\Phi,u}(t) = t^m$ where m is the number of eigenvalues of $\Phi''(u)$ which are less than zero. In the case when $\ker \Phi''(u)$ is one dimensional the following Morse Lemma is useful.
We write $H = H^- \oplus H^0 \oplus H^+$, where H^-, H^0 and H^+ are subspaces corresponding to negative, zero and positive eigenvalues of $\Phi''(u)$. The result of [3] asserts that there is a homomorphism D of a neighborhood of u into H and C^2 - map $\psi : H^0 \to R$ such that $D(u) = u$ and

$$\Phi(Dv) = \Phi(u) - 1/2\|v^- - u^-\|^2 + 1/2\|v^+ - u^+\|^2 + \psi(v^0 - u^0),$$

for all $v = v^- + v^0 + v^+ \in H^- \oplus H^0 \oplus H^+$ and $\|v - u\|$ small.
By $m^-(u)$ and $m^0(u)$ we denote the negative and zero Morse index of u i.e

$$m^-(u) := \dim H^- \text{ , } m^0 := \dim H^0 = \dim \ker \Phi''(u).$$

Proposition 2 *Let Φ be as above and $m^0(u) = 1$.*
Then

- *if u^0 is a local maximum of ψ then $P_{\Phi,u}(t) = t^{m^-(u)+1}$,*
- *if u^0 is a local minimum of ψ then $P_{\Phi,u}(t) = t^{m^-(u)}$,*
- *if u^0 is neither maximum or minimum of ψ then $P_{\Phi,u}(t) = 0$.*

In other words, if $\ker \Phi''(u)$ is one - dimensional then u contributes to Morse equation like a nondegenerate critical point or does not contribute at all (the third case). In this later case u is trivial.

Proof of Theorem 1. The proof is by induction. Let Σ^1 be a 1 - cube and let $C = [u, v]$ be an order interval corresponding to Σ^1. With a_i denoting number of critical points with Poincaré series t^i by applying Proposition 1 we have

$$2 + \sum_{i \geq 1} a_i t^i = 1 + (1+t)Q(t)$$

for some polynomial $Q(t) \in Z^+[t]$. Substituting $t = 1$ we deduce that total number $\sum_{i\geq 1} a_i$ of nontrivial critical points in $C^0 = C \setminus \{u, v\}$ is odd. Hence C contains at least 3 nontrivial critical points.
Now assume that result holds for any k - cube with $k \leq n-1$.
Let Σ^n be n - cube with corresponding order interval $C = [u, v]$. There are exactly $2n$, $(n-1)$ - subcubes of Σ^n. By $C_1, \ldots, C_{2n}$ we denote corresponding order - interval. By induction assumption, each C_i^0 contains the set S_i of nontrivial critical points and none of the elements of S_i is a local minimum of ψ. We claim that $S_i \cap S_j = \emptyset$, for $i \neq j$. This is obvious if $C_i \cap C_j = \emptyset$. Hence assume $C_i \cap C_j \neq \emptyset$, for some i, j. For simplicity we write $i = 1$, $j = 2$ and $C_1 = [u_1, v_1]$, $C_2 = [u_2, v_2]$.
It can be shown that there is an order - interval $C_{12} = [u_3, v_3]$ such that $C_{12} \subseteq C_1 \cap C_2$ and if u is a local minimum of Φ which is contained in $C_1 \cap C_2$ then $u \in C_{12}$. Furthermore, we can assume that either $u_1 = u_2 = u_3$, $v_3 \leq v_1$, v_2 and v_1, v_2 are noncomparable or $u_1 \ll u_3 = u_2 \ll v_3 = v_1 \ll v_2$. In the later case $C_{12} = C_1 \cap C_2$ and since in this case $S_1 \cap C_{12} = S_2 \cap C_{12} = \emptyset$ we have $S_1 \cap S_2 = \emptyset$.
In the first case assume there is $p \in S_1 \cap S_2$. Put $B := [p, v_1] \cap [p, v_2]$. Obviously B is convex, closed, int $B \neq \emptyset$ and the condition (Φ) implies that $K(B) \subseteq B$ and that Φ is bounded from below. Then there is $p_1 \in B$ such that $\Phi'(p_1) = 0$ and $\inf_B \Phi = \Phi(p_1)$. This can be shown by using a flow of $x' = -\Phi'(x)$ and previous remark. Moreover, since K is strongly order -

preserving $p_1 \ll v_1$ and $p_1 \ll v_2$.
In the case when $m^-(p) \geq 1$ by applying Krein - Rutman theorem we can show that there is $p_2 \in \text{int}B$ such that $\Phi(p) > \Phi(p_2)$. Hence p_1 is a local minimum of Φ and $p_1 \in C_{12}$. But since $p \ll p_1$, $p \in C_{12}$ which is impossible. The other possibility $m^-(p) = 0$ implies $m^0(p) = 1$ (otherwise p is a local minimum) and since p is nontrivial p^0 is either a local minimum or a local maximum of ψ (see Morse Lemma).
In the first case p must be a local minimum of Φ and in the second by applying again Krein - Rutman result and Morse Lemma we can find $p_2 \in B$ so that $\Phi(p) > \Phi(p_2)$. As before p_1 is a local minimum of Φ and this is impossible. Hence $S_1 \cap S_2 = \emptyset$.
Now let a_i be the number of nontrivial critical points in C whose Poincaré series are t^i. Then $a_0 = 2^n$ and by Morse equation

$$2^n + \sum_{i \geq 1} a_i t^i = 1 + (1+t) \sum_{i \geq 1} b_i t^i \quad \sum_{i \geq 1} b_i t^i \in Z^+[t].$$

Taking $t = 1$ we get

$$2^n + \sum_{i \geq 1} a_i = 1 + 2 \sum_{i \geq 1} b_i$$

and the total number $\sum_{i \geq 1} a_i$ of critical points in C is odd.
On the other hand, by the previous remarks the number of nontrivial critical points in $\bigcup_1^{2n} C_i$ is equal to $b := \sum_{k=0}^{n-1} \sum_{\Sigma_0} b_{\Sigma_0}$, where Σ_0 is a k - subcube of Σ^n and b_{Σ_0} is the number of nontrivial critical points in L^0 (L is an order - interval associated with Σ_0). But by the induction assumption b_{Σ_0} is odd for any subcube $\Sigma_0 \subseteq \Sigma^n$ and since for a given $k \in \{0, \ldots, n-1\}$ the number of all k - subcubes of Σ^n is $2^{n-k} \begin{pmatrix} n \\ k \end{pmatrix}$ we deduce that b is even. Hence there must be an odd number of nontrivial critical points in C^0.
Finally, since each order - interval L associated to k - subcube of Σ^n ($k = 0, 1, \ldots, n$) contains at least one nontrivial critical point in L^0 there are at least $\sum_{k=0}^{n} \begin{pmatrix} n \\ k \end{pmatrix} 2^{n-k} = 3^n$ critical points in C. □

Remark. In the case when Σ^n is a 1 - cube or 2 - cube we can show that the order - interval C corresponding to Σ^n must contain at least one critical point with the Poincaré polynomial t^n ($n = 1, 2$). In fact, if $n = 1$ then from

Morse equation we have

$$2 + a_1 t + a_2 t^2 + \cdots = (1 + b_1) + (b_1 + b_2)t + \cdots$$

where $\sum b_i t^i \in Z^+[t]$, and a_1 must be at least equal to 1.
Similarly, if $n = 2$ we have

$$4 + a_1 t + a_2 t^2 + \cdots = (1 + b_1) + (b_1 + b_2)t + (b_2 + b_3)t^2 + \cdots$$

and since $a_1 \geq 4$ (Σ^2 contains exactly four 1 -subcubes and each 1 - subcube contains at least one critical point with the Poincaré polynomial t) we conclude $a_2 \geq 1$.

As for the proof of Theorem 2, it follows from Theorem 1. If Σ^n is an n - cube in Σ and $C = [u, v]$ is a corresponding order - interval then there is at least one nontrivial critical point u_0 in C^0. We should only prove that if Σ^n, Σ^k are different cubes in Σ and C_0, C_1 are associated order - intervals then the nontrivial solutions in C_0^0 and C_1^0 are different. But this follows from consideration similar to the one in the proof of Theorem 1 and we omit details.

3 Application.

In this section we prove Theorem 3 by applying our abstract results.
The solutions of Equation 1 are the steady states of a related parabolic problem $\epsilon u'_t = u_{xx} + u(1 - u)(u - a(t))$, $u'(0) = u'(1) = 0$ which serves as a simple model for genetic evolution.
Recently problem (1) has been studied by Angenent, Mallet - Paret and Peletier in [1]. They gave a complete classification of all stable solutions of (1).
Since we will need their results we describe them briefly.
For definiteness we assume that $a(0) > 1/2$. Let $Z = \{t \in [0, 1];\ a(t) = 1/2\}$. Then the main result of [1] says that there is $\epsilon_0 > 0$ such that for any $\epsilon \in (0, \epsilon_0)$ and any $\{t_1 < \cdots < t_m\} \subseteq Z$ there is a stable solution of (1) and $a'(t_i)u'(t_i) < 0$, for each $i = 1, \ldots, m$, u is monotone in a small neighborhood of each t_i and away from t_i, $u(t)$ is close either to 0 or 1. Moreover, all stable solutions are obtained in this way.

Here the stability of u means that the principal eigenvalue μ of the linearized problem

$$\epsilon v''(t) + f_\xi(t, u(t))v(t) = \lambda v(t)$$

is nonpositive.
In [1] it is shown that the principal eigenvalue of the above linearized problem at u which is as above must be necessarily strictly less than zero.
We are interested in the least number of solutions of (1) for $\epsilon \le \epsilon_0$. For the sake of simplicity we take $\epsilon = 1$.
We take

$$H = H^1 = \{u\ ;\ u \text{ is absolutely continuous and } u' \in L^2[0,1]\}$$

with the inner product defined by

$$(u,v)_\lambda := \int_0^1 u'v' + \lambda \int_0^1 uv\ ,\ \lambda > 0.$$

Here $\lambda > 0$ is chosen so large that if $f_\lambda(t,\xi) := f(t,\xi) + \lambda\xi$, where $f(t,\xi) := \xi(1-\xi)(\xi - a(t))$ then $f_\lambda(t,\cdot)$ is increasing and $f_{\lambda,\xi}(t,\xi) > 0$ for $(t,\xi) \in [0,1]^2$.
Let $P = \{u \in H\ ;\ u(t) \ge 0,\ t \in [0,1]\}$.
Then (H, P) is an ordered Hilbert space and P is a cone with nonempty interior.
We define

$$\Phi(u) := 1/2\|u\|_\lambda^2 - \int_0^1 F_\lambda(t, u(t))dt, \text{ for } u \in H,$$

where $F_\lambda(t,\xi) = \int_0^\xi f_\lambda(t,s)ds$.
We have

$$\begin{aligned}
\Phi'(u)h &= (u,h)_\lambda - \int_0^1 f_\lambda(t,u(t))dt \\
&= (u - K(u), h)_\lambda \\
(\Phi''(u)h, v)_\lambda &= (h,v)_\lambda - \int_0^1 f_{\lambda,\xi}(t,u(t))v(t)h(t)dt \\
&= (h - K'(u)h, v)_\lambda
\end{aligned}$$

for $u, v, h \in H$.
The critical points of Φ are classical solutions of (1). It is rather a straightforward exercise to show that Φ satisfies condition (Φ) and that dim ker

$\Phi'' \leq 1$ for any critical point u of Φ. Furthermore, the stability of a solution u means $(\Phi''(u)h, h)_\lambda \geq 0$, for $h \in H$ and if u is a stable solution of (1) then the negativity of the principal eigenvalue implies

$$\inf_{h \in H \setminus \{0\}, \|h\|_\lambda = 1} (\Phi''(u)h, h)_\lambda > 0.$$

Hence if u is stable solution of (1) then it must be a local minimum of Φ. On the other hand if u is a local minimum of Φ then u is a stable solution of (1). Thus the set of all stable solutions coincides with the set of all local minima of Φ which we denote Σ.
According to Theorem 2 the least number of solutions of (1) is the same as the number of all cubes which appear in the graph (Σ, Γ) where Γ is the set of edges defined in Introduction. To find this number we use special form of elements of Σ.

Assume that $Z = \{t_1 < \ldots < t_2\}$ and $t_0 = 0, t_{k+1} = 1$. With each $u \in \Sigma$ we associate a vector of $k+1$ - numbers $\alpha(u) = (\alpha_1, \ldots, \alpha_{k+1})$ in the following way :

$$\begin{aligned} \alpha_i &= 0 \text{ if } u(\frac{t_{i-1} + t_i}{2}) < 1/2 \text{ and} \\ \alpha_i &= 1 \text{ if } u(\frac{t_{i-1} + t_i}{2}) > 1/2. \end{aligned}$$

From $a'(t_i)u'(t_i) < 0$ (if $u \in \Sigma$) it follows that $\alpha(u) = (\alpha_1, \ldots, \alpha_n)$ must satisfy :

- $\alpha_i = 1$ for i odd implies $\alpha_{i+1} = 0$ and
- $\alpha_i = 0$ for i even implies $\alpha_{i+1} = 0$.

We denote by $\mathcal{A} = \mathcal{A}(k)$ the set of all vectors $(\alpha_1, \ldots, \alpha_{k+1})$ ($\alpha_i \in \{0, 1\}$) which have the above property and introduce ordering on $\mathcal{A}$ by

$$(\alpha_1, \ldots, \alpha_{k+1}) \leq (\beta_1, \ldots, \beta k + 1) \Leftrightarrow \alpha_i \leq \beta_i.$$

Then we have

Lemma 1 $\alpha : \Sigma \longrightarrow \mathcal{A}$ *is a bijection and* $u \ll v \Leftrightarrow \alpha(u) < \alpha(v)$.

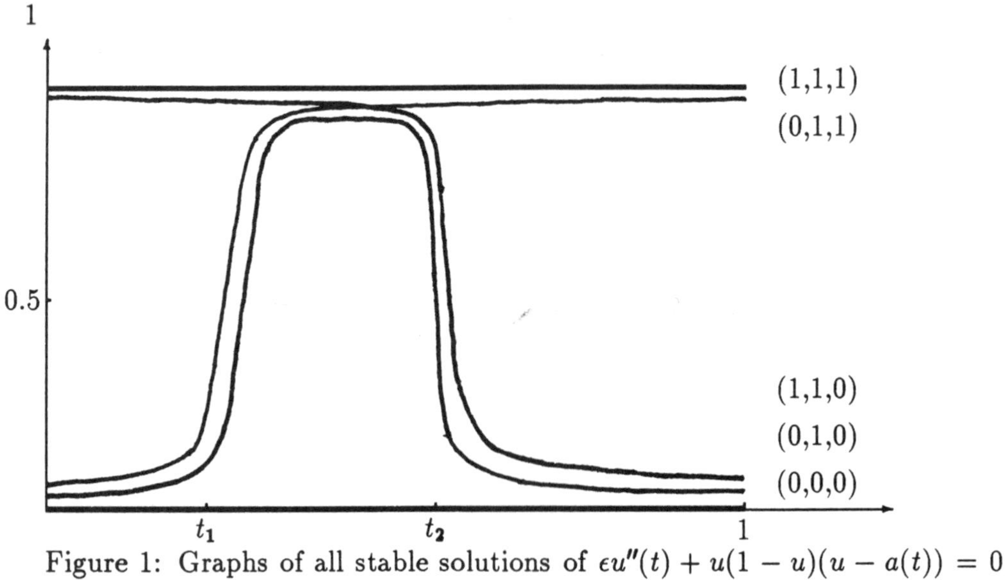

Figure 1: Graphs of all stable solutions of $\epsilon u''(t) + u(1-u)(u-a(t)) = 0$ and their codes when $a(t) - 1/2$ has 2 zeros.

If $\Gamma' \subseteq \mathcal{A} \times \mathcal{A}$ is the set defined by

$$\{(\alpha, \beta) \; ; \; \alpha, \beta \in \mathcal{A}, \alpha \leq \beta \text{ and } \sum_{i=1}^{k+1} |\beta_i - \alpha_i| = 1\}$$

then Lemma 1 says that (Σ, Γ) and $(\mathcal{A}, \Gamma')$ as graphs are isomorphic.

Let $N_l(k)$ be the number of all l - cubes in the graph $(\mathcal{A}, \Gamma')$, $\mathcal{A} = \mathcal{A}(k)$. Then $N_l(k) = 0$ if $l - 1 \geq k > 0$ (since $|\mathcal{A}| < 2^{k+1}$ and the points of $\mathcal{A}(k)$ cannot form l - cubes if $l - 1 \geq k > 0$).

It is proved in [4] that the numbers $N_l(k)$ satisfy following recurrent formulas

$$N_0(0) = 2, N_0(1) = 3, N_1(0) = 1, N_1(1) = 3$$
$$N_0(k) = N_0(k-1) + N_0(k-2)$$
$$N_l(k) = N_l(k-1) + N_l(k-2) + N_{l-1}(k-2) \text{ for } l \in Z^+, k \geq 2$$

Using this relations the number $N(k)$ of all cubes must satisfy

$$\begin{aligned} \sum_{l \geq 1} N_l(k) &= \sum_{l \geq 1} N_l(k-1) + \sum_{l \geq 1} N_l(k-2) + \sum_{l \geq 1} N_{l-1}(k-2) \\ &= \sum_{l \geq 1} N_l(k-1) + \sum_{l \geq 1} N_l(k-2) + \sum_{l \geq 0} N_l(k-2) \end{aligned}$$

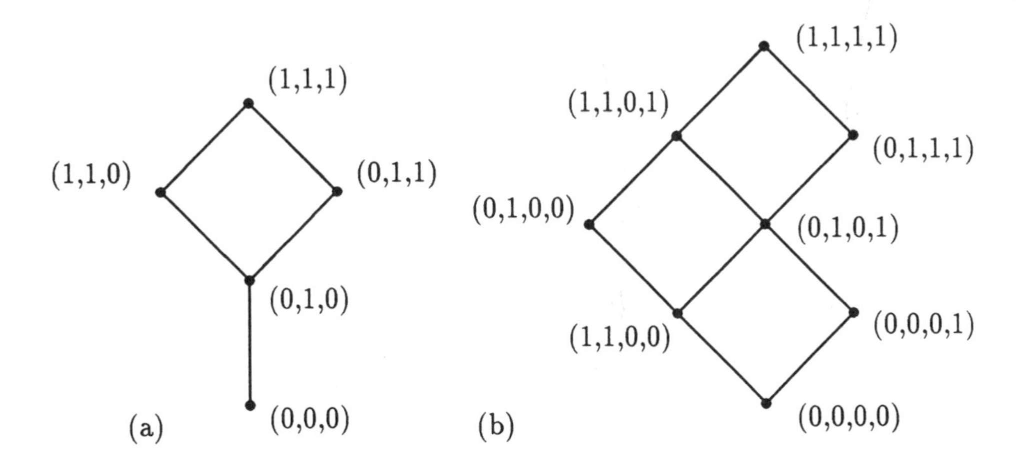

Figure 2: An ordering of all local minima of Φ when the function $a(t) - 1/2$ has: (a) 2 zeros and (b) 3 zeros.

But since $N_0(k) = N_0(k-1) + N_0(k-2)$ we deduce $N(k) = 2N(k-2) + N(k-1)$.
Now from that it follows easily by induction that

$$N(k) = \frac{2^k + 2(-1)^k}{3} N(0) + \frac{2^k + (-1)^{k+1}}{3} N(1) \text{ for } k \geq 2$$

and since $N(0) = 3$ and $N(1) = 5$ we get

$$N(k) = \frac{2^{k+3} + (-1)^k}{3} \text{ for } k \geq 2$$

and the proof is completed.

It is interesting to compare $N_0(k)$ and $N(k)$ (i.e number of all stable solutions and the least number of all solutions) for some values of k.

k	$N_0(k)$	$N(k)$
0	2	3
1	3	5
5	21	85
10	233	2 731
15	2 584	87 381
20	28 657	2 796 203
25	317 811	89 478 485

References

[1] Angenent, S.B.; Mallet - Paret, J., Peletier, L.A. ; Stable transition layers in semlinear boundary value problem, J. Diff. Equat. 62(1986), 427 - 442.

[2] Hofer, H.; Variational and topological methods in partially ordered Hilbert spaces, Math. Ann. 261(1982), 493 - 5 14.

[3] Hofer, H.; The topological degree at a critical point of mountain - pass type, Proceeding of Symposia in Pure Math., Vol 45(1986), 501 - 509.

[4] Wysocki, K.; Multiple critical points for variational problems on ordered Hilbert spaces, submitted.

Index